THE NUCLEAR QUESTION

Also by Ann E. Weiss

WHAT'S THAT YOU SAID?
  How Words Change

? THE NUCLEAR QUESTION

by Ann E. Weiss

Illustrated with photographs

HARCOURT BRACE JOVANOVICH
New York and London

ACKNOWLEDGMENT. The author gratefully acknowledges the information and help given her by Donald Vigue of the Maine Yankee Atomic Power Company and by Jeanne Cook of the Nuclear Regulatory Commission.

PHOTO CREDITS. Atomic Industrial Forum, page 21; Maine Yankee, pages 8, 22, 24, 42, 44, 54, 56–57, 68, 78, 109; NRC, pages 30, 49, 50, 52, 63, 72–73, 87, 94, 151; Malcolm E. Weiss, pages 115, 138

Copyright © 1981 by Ann Weiss

*All rights reserved. No part of
this publication may be reproduced or
transmitted in any form or by any means,
electronic or mechanical, including photocopy,
recording, or any information storage and
retrieval system, without permission
in writing from the publisher.*

*Requests for permission to make copies of
any part of the work should be mailed to:
Permissions, Harcourt Brace Jovanovich, Inc.,
757 Third Avenue, New York, N.Y. 10017*

Printed in the United States of America

Library of Congress Cataloging in Publication Data

*Weiss, Ann E., 1943–
The nuclear question.
Bibliography: p.
Includes index.
Summary: Discusses the development of nuclear power, its benefits,
dangers, and future, and the controversy surrounding it.
  1. Atomic energy—Juvenile literature.   2. Atomic power-plants—
Juvenile literature. [1. Atomic energy. 2. Atomic power plants] I.
Title.
TK9148.W44   333.97′24   80-8806
ISBN 0-15-257596-0   AACR2*

BCDE   *First edition*

CONTENTS

1. THE WORLD SET FREE? — 9
2. INTO THE ATOM — 13
3. THE PEACEFUL ATOM — 20
4. THE WHYS AND HOWS OF RADIATION — 31
5. THE BROKEN CYCLE — 47
6. SAFETY SYSTEMS AND AN ACCIDENTAL TEST — 66
7. OTHER ACCIDENTS, OTHER THREATS — 83
8. REGULATION OR PROMOTION? — 92

9   PAYING THE NUCLEAR BILL   104

10   THE BATTLE OF WORDS   113

11   MISDIRECTION, MISINFORMATION, AND OFFICIAL DECEPTION   122

12   OUR ENERGY FUTURE   134

    BIBLIOGRAPHY   153

    INDEX   156

THE NUCLEAR QUESTION

The huge dome of a nuclear power plant overlooks a scenic spot on the Maine coast. High tension wires carry electricity from the plant to homes, schools, factories, stores, and offices around New England. Is nuclear power a safe, clean, and cheap means of producing electricity? Or does it represent a deadly danger—today and far into the future?

1
THE WORLD SET FREE?

It was a world like our world, in a time like our time.

Man had used the earth's energy resources recklessly. Oil supplies were dwindling. So were coal reserves. Millions of acres of forest land had been swept away by loggers. Miners had exhausted once rich lodes of tin and copper deposits. Even water resources were affected by man's greedy energy consumption. Lakes and rivers had been dammed up, lowering underground water tables and producing periodic droughts.

This picture of twentieth-century life was painted nearly seventy-five years ago by the English writer H. G. Wells. In a science-fiction novel called *The World Set Free*, Wells presents a grim vision of man's struggle to exist on a planet rapidly running out of natural energy sources. In many ways, Wells's fictional world resembles our real world of today—a world in which growing energy shortages pose a serious threat to our way of life.

But there were, in Wells's novel, people who believed that a whole new kind of energy lay waiting to serve mankind. These people were scientists, and they were confident that it would one day be possible to turn every bit of matter in the world into a source of power. How? By breaking apart the individual atoms of which matter is composed. Wells's fictional scientists were convinced that an enormous amount of energy was stored in the nucleus—the core—of every atom in the universe. Splitting the atom would release that energy.

As the novel opens, Professor Rufus of Edinburgh, Scotland, is explaining the atomic theory to his students:

> "And we know now that the atom, that once we thought hard and impenetrable, and indivisible and final and—lifeless—lifeless, is really a reservoir of immense energy. . . . We thought of the atoms as we thought of bricks . . . and behold! these bricks are boxes, treasure boxes, boxes full of the intensest force. . . ."

When Wells wrote *The World Set Free,* he was one of only a very few people who believed that nuclear energy existed and that scientists would one day discover how to release it and put it to work. In his book, Wells set the date of the discovery as 1933.

The fictional breakthrough came in the London laboratory of a physicist named Holsten. Wells describes Holsten's experiment:

> He set up atomic disintegration in a minute particle of bismuth, it exploded with great violence. . . . But the thing was done,—at the cost of a blistered chest and an injured finger, and from the moment when the invisible speck of bismuth flashed into riving and rending

energy, Holsten knew that he had opened a way for mankind . . . to worlds of limitless power.

Holsten had. Within ten years, Wells writes, people were flying their own private nuclear-powered helicopters. By the 1950s, towns and cities were brightly lit by electricity produced in nuclear power plants. Atomic automobiles—"light and clean and shimmering"—ran at a cost of 1/37 of a penny per mile. Energy was so plentiful and so inexpensive that entire sections of cities could be torn down and rebuilt. The whole world was running on nuclear power produced by machines whose waste product was—gold. Here, indeed, was the fulfillment of Professor Rufus's dream: " . . . I see the desert continents transformed, the poles no longer wildernesses of ice, the whole world once more Eden. I see the power of man reach out among the stars."

All this, Holsten foresaw on that day in 1933. After his great discovery, Wells says, the scientist went home, worried that he might not sleep because of the pain in his hand and chest, and his elation at what he had achieved. "Slept like a child," the contented Holsten wrote in his diary.

The World Set Free was published in 1913, and it must have caused many people to wonder whether nuclear power really were feasible. When a Hungarian biologist named Leo Szilard read the book in 1932, he was at once caught up in Professor Rufus's dream. As Szilard wrote years later, he decided then and there that, "If I wanted to contribute something to save mankind, then I would probably go into nuclear physics, because only through the liberation of atomic energy could we . . . enable man not only to leave the Earth but to leave the Solar System."

Szilard did go into nuclear physics. One day in 1933—the very year of Holsten's discovery—there flashed into Szilard's mind an idea for a method of getting atoms to give off their nuclear energy in a usable form.

Six years later the idea had become reality, and on March 3, 1939, Szilard and other nuclear physicists performed a key experiment at New York City's Columbia University. Like Holsten, Szilard recorded the event and his feelings about it:

> Everything was ready and all we had to do was to turn a switch, lean back, and watch the screen of a television tube. If flashes of light appeared on the screen, that would mean [that the experiment was working] and this in turn would mean that the large-scale liberation of atomic energy was just around the corner. We turned the switch and saw the flashes.

The experiment was going well. But Szilard's ideas had changed since 1932, and he was no longer certain that nuclear energy would save humanity or carry mankind to the stars. He knew that Wells's concept of nuclear energy as a universal power source was mistaken; it could not be used to run machines like cars and airplanes. Nuclear power's civilian use would be limited to producing electric power in central generating stations. Szilard knew, too, that the energy of the atom could be used to build bombs with a hideous capacity for destruction, and he was aware of the deadliness of the radioactivity given off in a nuclear reaction.

So Szilard was subdued as he observed the experiment: "We watched . . . for a little while and then we switched everything off and went home. That night there was very little doubt in my mind that the world was headed for grief."

A world headed for grief? Or a world set free? The nuclear power debate was just beginning on that March evening. At first, the debate involved only a handful of scientists on either side. Today, the debate continues, given new importance by the March 1979 accident at the Three Mile Island nuclear plant in Middletown, Pennsylvania. That debate now involves every one of us—girl or boy, man or woman—in every part of the country, and in every nation of the world.

2 INTO THE ATOM

Nuclear power starts with atoms.

The word "atom" comes from the Greek word *atomos.* In about 400 B.C., the Greek philosopher Democritus suggested that if matter were divided into smaller and smaller pieces, each bit would finally be so tiny that it could not be divided any more. Each bit would be *atomos*—something that cannot be cut. Each would be a solid ball of matter that could not be changed in any way.

For over two thousand years, no one gave much thought to this theory. Then, in the late 1800s, scientists began making one discovery after another that challenged Democritus's ideas.

In the 1890s, the English physicist J. J. Thompson found that atoms are not solid balls after all. Instead, he learned, each one is made up of individual particles. These particles are electrical in nature, and can interact with one another and with particles in other atoms. Those that Thompson discovered carry a negative electrical charge. They came to be called electrons.

A few years later, another scientist, Ernest Rutherford, showed that electrons are not the only particles in atoms. His experiments demonstrated that atoms also contain particles with a positive charge. Called protons, these particles are clustered together to form a nucleus in each atom. The positive charge of the nucleus attracts the atom's electrons, and keeps them whirling endlessly around it, rather like planets circling the sun. Since the number of protons in each atom equals its number of electrons, the atom as a whole has no electrical charge.

A third kind of subatomic particle—the neutron—was discovered in the 1920s. As the name suggests, neutrons have no electrical charge. They are neutral. An atom's neutrons are found in the nucleus along with its protons.

Scientists were learning other facts about atoms. They knew that the atoms of each chemical element are different from the atoms of every other element. One difference lies in the number of protons in each atom of a particular element. All hydrogen atoms have 1 proton, for instance. Helium atoms always have 2 protons. Carbon atoms have 6. Uranium atoms have 92. Or you can state it another way. *Any* atom with 6 protons in its nucleus is a carbon atom. *Any* atom with 92 protons is a uranium atom.

But scientists also knew that atoms of the same element sometimes have different numbers of neutrons in their nuclei. Atoms of the same element with varying numbers of neutrons are called "isotopes" of that element.

Take the element hydrogen, for instance. Most hydrogen atoms consist of one proton with one electron circling it. But the nuclei of a few hydrogen atoms also contain one neutron. These atoms are larger and heavier than ordinary hydrogen atoms, so they are called "heavy hydrogen." A very small number of hydrogen atoms have one proton, one electron, and two neutrons. Called "tritium," these isotopes weigh even more than heavy hydrogen. Nearly all the elements have more than one isotope, and some have several. Uranium has at least fourteen.

Another fact that scientists learned about atoms was that they are not forever changeless and indivisible. One early demonstration of that came about by accident.

In 1896, the French scientist Antoine Henri Becquerel placed some uranium on a photographic plate that was wrapped in heavy paper, and put the plate in a drawer. When he took it out a few days later, he discovered a dark smudge on the plate. It marked the exact spot where the uranium had rested.

Becquerel considered what had happened. Only some kind of energy could have made the mark. But the only thing in the drawer besides the plate had been the uranium. Could *that* have been the source of the energy?

Becquerel decided that it was. He called the energy given off by the uranium "radioactivity"—ray action. (Our words "ray" and "radio" come from the same Latin root.)

Soon other scientists were investigating radioactivity. Chemist Marie Curie and her physicist husband, Pierre, set out to find out whether there were other radioactive elements in addition to uranium. After four years of research at their Paris laboratory, the Curies had found two, radium and polonium. Later scientists found others. They, like the Curies, observed that radioactive elements give off energy in the form of light and heat.

What's more, the Curies learned that besides being radioactive itself, an element like radium can make the things around it radioactive, too. The element's container, the table on which it sits, even the surrounding air, all glow with radioactive energy.

Experiments by the Curies and others uncovered more information about radioactivity. Ernest Rutherford found that radioactive elements give off radiation in three different forms. Rutherford named them after the first three letters of the Greek alphabet, *alpha, beta,* and *gamma.*

Alpha rays, or particles, consist of two protons and two neutrons. They have a positive charge. Alpha particles are the least penetrating form of radioactivity, and can be blocked by a piece of paper or by unbroken human skin.

Beta rays, or particles, have a negative charge. They are smaller and lighter than alpha particles, and they move faster and more energetically. They can go through several sheets of paper or travel deep into the human body.

Most penetrating of all are gamma rays. They easily go through paper or wood and can penetrate human bone and tissue. It takes a lead shield, half an inch (1.27 cm) or more thick, to stop gamma rays, which are very much like the X rays used in medicine, industry, and other fields.

Another of Rutherford's findings was that as an element gives off radioactive alpha, beta, or gamma rays, it changes into other elements. As uranium emits radioactivity, for example, it changes by stages into radium. Then it becomes polonium, and finally, a nonradioactive substance—lead. This process is called transmutation of elements, radioactive breakdown, or radioactive decay.

Radioactive decay takes place at a regular rate. In any sample of radium, no matter how large or small it is, half of the atoms will have decayed into atoms of lead at the end of 1,660 years. Of the remaining radium atoms, half will have become lead at the end of another 1,660 years. Now only one-fourth of the original radium is left. Half of that will break down over the next 1,660 years, and so on. The period of time it takes for half of a radioactive element to decay is called that element's "half-life." Some radioactive materials have very short half-lives. The half-life of radioactive fluorine is eleven seconds. The half-life of uranium, on the other hand, is 4.5 billion years.

By the 1920s, scientists knew quite a bit about radiation and the atom. But so far, their knowledge did not add up to any practical means of capturing nuclear power to meet man's energy needs. That was because scientists did not know how to release nuclear energy except through radioactive decay. And radioactive decay does not give off enough energy fast enough to run machines or produce electricity. If people were to benefit from nuclear power, they would have to learn to create more intense nuclear reactions. But how?

Gradually scientists evolved an idea. Perhaps if they aimed a stream of particles at atomic nuclei, the particles would crash into the nuclei and break them apart. The breaking apart would release a large part of their nuclear energy all at once.

At first, scientists used streams of alpha particles. That didn't work. The reason: alpha particles are positive, and so are the nuclei of atoms. Similar electrical charges push each other apart, as opposite ones attract. Most of the alpha particles simply bounced off the nuclei.

Then, in the 1920s, came the discovery of the neutron—the perfect "bullet" for splitting atoms.

Splitting the Atom

One of the first researchers to use the neutron bullet was an Italian physicist named Enrico Fermi. Working at the University of Rome, Fermi methodically bombarded the nuclei of one element after another with neutrons. In some cases, the bombardment had no effect. But Fermi found he could get the atoms of many elements to break apart and give off energy.

Fermi also discovered that he could get more energy from an element if he placed a barrier of water between the bombarding neutrons and the element being bombarded. This puzzled him until he realized that the barrier was slowing down, or moderating, the neutrons. Slow neutrons are more likely to smash into a nucleus than fast neutrons are. Fast neutrons go through a nucleus too quickly to change it.

At about the time Fermi was starting his work in Rome, Leo Szilard, in London, was reading *The World Set Free* and coming up with his idea for getting usable energy from atomic nuclei.

Szilard's idea was simple. All he had to do was to find an element whose nuclei, when broken apart by *one* neutron, would emit *two* neutrons. Each of the newly freed neutrons would be available to break apart another nucleus, and each of those two nuclei would emit two more neutrons. Now there would be four neutrons. If each of those split another nucleus, there would be eight. Then sixteen, then thirty-two, and so on and on, in a nuclear chain reaction—a reaction that, once started, keeps itself going.

By the late 1930s, Szilard and Fermi were working together at Columbia University. There, on March 3, 1939, flashes of light on a TV screen showed that the chain reaction worked.

Actually, Szilard and Fermi later learned that they were not the first to get uranium to give off enough extra neutrons for a chain reaction. Two German scientists had done the same thing a year earlier. But the Germans had trouble getting news of their accomplishment to the United States because the German government was keeping its nuclear research a top secret. With Europe on the brink of World War II, Germany wanted to be the first country capable of developing nuclear weapons. One of the people who helped get word of the German discovery to the outside world was the Austrian physicist Lise Meitner. She was also the one who first used the word "fission" to describe the breaking up of atoms. ("Fission" comes from a Latin word that means "to split.")

Like the Germans, United States officials were preparing for war—and eyeing the possibility of a nuclear bomb. By 1941, Fermi's lab had been moved to Chicago, and he was receiving large shipments of uranium and other materials.

Fermi's new laboratory was an unused squash court at the University of Chicago. There, he and his assistants erected a great pile of graphite blocks. The graphite was highly refined; it was much purer than the graphite we use in ordinary "lead" pencils. It would act as a moderator, slowing down neutrons—and thereby speeding up fission.

Among the graphite blocks were chunks of uranium. This was the "fuel" within which Fermi hoped to sustain the nuclear chain reaction.

On December 2, 1942, the "atomic pile" was ready for its big test. A young man stood beside the pile, his hand on a rod that reached deep within the graphite and uranium. The rod was made of cadmium, an element that can absorb neutrons without the nuclei of its own atoms breaking up. As long as the cadmium rod remained inside the pile, it would trap too many neutrons to let a chain reaction start. But when the rod was taken out, neutrons would begin to crash into the uranium nuclei. If all went well, each collision would release more neutrons. The chain reaction would be under way.

Fermi gave the signal, and the young man pulled the rod a little way out of the pile. Geiger counters, used to measure radioactive emissions, began to click. The rod came farther out. The counters clicked faster. The pile was producing energy!

Now Fermi gave the order to pull the rod out even more. The clicks became a steady stream of noise. The chain reaction sustained itself for over thirty minutes.

Would pushing the cadmium control rod back into the pile stop it? Again, Fermi signaled. Softer and softer became the sound of the Geiger counters. At last they were still. The experiment was over. Mankind had unleashed the energy of the atomic nucleus.

On August 6, 1945, the first public demonstration of that energy took place. The United States dropped a uranium fission bomb on Hiroshima, Japan. That one bomb destroyed an area of 4.5 square miles. It killed more than 90,000 men, women, and children, and injured as many more. Three days later, another atomic bomb—this one fueled with the element plutonium—fell on the city of Nagasaki. Again, the damage was enormous. On August 14, Japan surrendered to the United States. World War II was over.

The nuclear age had begun.

3
THE PEACEFUL ATOM

World War II was over, but research into nuclear weapons continued. In 1948, the United States exploded a bomb six times more powerful than the one that had destroyed Nagasaki. The next year, the Soviet Union set off its first atomic bomb. Other countries, including England, France, and China, followed suit. Today, the world's great powers and some smaller nations, too, are members of the so-called "nuclear club."

But nuclear energy can do more than destroy, and much of the postwar research has been aimed at finding peaceful uses for it. This research has led to new, precise ways of diagnosing illnesses. It has given doctors powerful tools for fighting cancer and other diseases. It has brought about advances in industry and agriculture. But perhaps the most important peaceful use of nuclear energy—and the most controversial one today—is in generating electricity.

In many ways, a nuclear power plant is no different from an electrical generating plant that burns a fossil fuel, such as coal, oil, or natural gas. Whatever the fuel, its job is to give off enough heat to produce steam at high pressure. The steam turns the propellerlike blades of a huge turbine. The turbine spins the shaft of a generator. That produces an electrical current. The electricity is carried over wires into our homes, schools, offices,

How a typical American nuclear plant works. Steam heated by the energy of fission turns the turbine, enabling a generator to produce electricity.

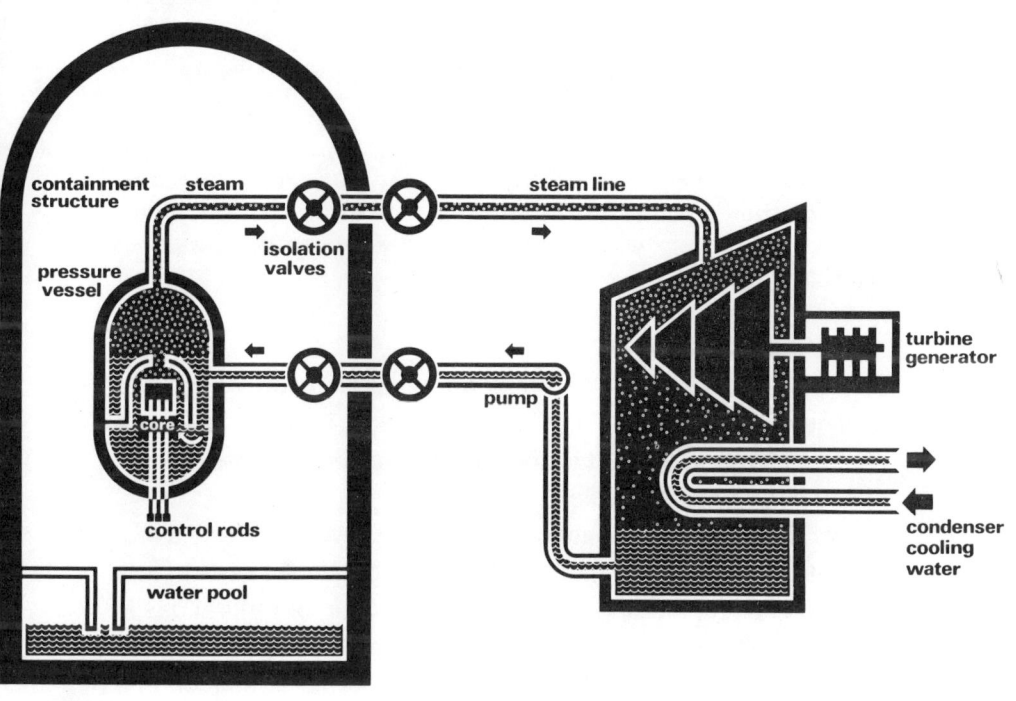

At a nuclear plant, steam-powered turbines spin the shaft of a generator, producing an electrical current. Electricity is generated the same way in fossil-fuel plants. But in a nuclear plant, the energy that produces the steam is supplied by fissioning uranium.

and factories. Meanwhile, the steam cools and condenses into its liquid form—water. It is pumped back into the pressure chamber, reheated, and converted into pressurized steam again, in an endless cycle.

The main difference between a nuclear power plant and a fossil fuel plant lies in the method used to produce the steam. In a fossil fuel plant, heating occurs when coal, oil, or natural gas is burned in a furnace—a larger and more elaborate version of the kind of furnace that probably warms your home or school. In a typical nuclear plant, the fuel is uranium. Uranium doesn't burn, though. It fissions. The nuclei of its atoms split, producing energy in the form of heat. The heat makes steam, which is passed through a series of metal coils. As the steam goes through the coils, it gives up its heat to a second batch of pressurized steam. This second batch of steam turns the turbine.

Fission takes place in a nuclear reactor—a sophisticated atomic pile of the type that Enrico Fermi built at the University of Chicago. The fuel in Fermi's pile was chunks of uranium. In a modern reactor, half-inch long pellets of uranium are packed into 12- or 14-foot (3.7- or 4.3-meter) tubes made of an alloy of the metal zirconium. About 50,000 zircalloy fuel rods make up the reactor core.

Fermi used graphite as a moderator; a typical American reactor uses water. Fermi controlled the chain reaction in his pile with a cadmium rod. In a present-day reactor, the control rods are likely to be made of boron, another effective neutron trap. To control the reaction further, the core may be flooded with a solution of boric acid.

A nuclear reactor works the way Fermi's pile did. With the control rods in place in the core, a chain reaction cannot begin. When the plant operators want the uranium to "go critical"— nuclear lingo for starting a chain reaction—they activate machinery that pulls the control rods away from the core. Once that is done, a single free neutron is enough to set off the reaction.

As the reaction continues, the moderator slows the neutrons down enough to ensure that they will continually split more uranium atoms. At the same time, the moderator acts as a coolant. It keeps the overall core temperature at about 580 degrees Fahrenheit (300 degrees Celsius), normal operating temperature for a nuclear reactor. Since the temperature at spots inside the fuel rods may be as high as 2,000 degrees Fahrenheit (1,100 Celsius), enormous amounts of coolant are continually needed to keep the core temperature at the proper level. Nuclear power plants are constructed near natural bodies of

The Peaceful Atom

The large cylindrical reactor head sits atop the reactor core. This photo was taken when the plant was not in operation. When fission is taking place, the reactor vessel is flooded with a boric acid solution.

water—lakes, rivers, or bays—so that cooling water can be pumped directly into and out of its natural source.

The steam produced by the heat of fission drives the turbine, generating electricity. Now the nuclear plant is "on line," sending electric power to its customers. When the plant must be shut down, its operators lower the control rods all the way back into the core. That brings the chain reaction to a standstill. The core cools, and steam is no longer produced.

Uranium vs. Fossil Fuels

The environmental effects of nuclear fuel differ from those of other fuels. The biggest difference involves nuclear radiation, which we'll look at in detail in the next chapter. But there are other differences in the effects that nuclear and nonnuclear fuels have on the environment.

Burning any fossil fuel makes soot and smoke that dirty the air. Coal is a particularly notorious polluter. It gives off more sulfur dioxide gas, which is harmful to animal and human life, than is allowed under the nation's clean air laws. Only by installing expensive antipollution devices can the pollution be cut down to safer legal levels. Oil and natural gas are less polluting than coal, but they, too, can dirty our atmosphere.

Not so uranium. Fission does not produce smoke or soot. Uranium is free of other fossil fuel hazards as well. It doesn't spill from ships, creating huge oil slicks on the ocean surface. Nor does it leak from pipelines to sink slowly into the earth.

However, nuclear plants do present pollution problems. One is the thermal pollution of nearby bodies of water. At a typical nuclear plant, only one third of the heat produced by fission actually heats steam to spin a turbine. The other two thirds is waste heat, which must be gotten rid of. In many cases, the warm water is simply pumped back into the natural body from which it came, where it kills plant and animal life. Thermal pollution can also bring about the rapid growth of certain life forms that may alter the balance of nature and endanger various other species. A separate but related problem is that fish eggs and larvae are often sucked up by the pumps and passed through the cooling system. This alone can kill up to three quarters of the fish at the site of a plant.

Environmental laws in some states require the construction of huge towers to cool the water used for cooling the plant. The water in these cooling towers is recycled—used over and over to keep plant temperatures at normal operating levels. The use of cooling towers reduces thermal pollution, but it can lead to other problems. As the water cools in the tower, as much as 1 percent of it may evaporate into the air. This can create foggy conditions in summer and icy ones in winter. It also tends to drain the source of the water, which can be particularly serious in a dry climate or during a drought.

Another difference between fossil fuel and uranium plants is that the former use fuel sources that are becoming scarce—and expensive. In the past hundred years, we have used up oil and gas deposits that formed over hundreds of millions of years. And until very recently our demand for energy had been growing rapidly. By one estimate, Americans will consume as much energy during the next twenty years as we used from the time of the Revolutionary War to the present day. Other estimates, those that take account of a leveling off of energy demands in the late 1970s, predict a somewhat lower rate of consumption between now and the year 2000.

Where will we get the fossil fuels to supply energy for the future? Increasingly, from foreign countries. In 1970, the United States itself produced more than 9 million barrels of oil each day. (One barrel holds 42 gallons of oil.) Nine years later, domestic production had dropped to 8 million barrels a day. During the same time, the amount of oil the United States imported from other nations rose sharply. Similarly, the United

States imports a greater percentage of its natural gas today than it did in the early 1970s.

Will other nations always be willing to sell their fossil fuels to us? The United States buys much of its oil from countries of the Middle East. Political tensions between this country and the Arab oil producers could lead to a reduction of the amount of oil available to American buyers. During the Arab oil embargo in 1974, millions of Americans were unable to buy gasoline for their cars and worried about whether they could continue to heat their homes. Another embargo is a real possibility. A war in the Middle East—and there have been several wars there in recent decades—could cut off the flow of oil altogether.

Not only will natural gas and oil become scarcer as the twenty-first century approaches, they will become more expensive. In 1967, imported oil cost about $1.80 a barrel. By mid-1980, it cost up to $37 a barrel.

High prices and uncertain supplies mean that the cost of anything that uses oil or natural gas will rise, too. This is borne out by figures from the Atomic Industrial Forum (AIF), a trade association of nuclear power companies, reactor designers and manufacturers, and others in the nuclear power industry. In 1976, according to the AIF, it cost 3.5 cents to produce one kilowatt-hour of electricity in an oil-burning plant. (One kilowatt-hour is the amount of energy it would take to keep a 1,000-watt bulb lit for one hour.) The next year, the cost rose to 3.9 cents per kilowatt-hour, and by 1978, it rose to 4 cents. Producing electricity in coal-fired plants also became more costly over the three-year period. By contrast, says the AIF, the cost of producing one kilowatt-hour in an existing nuclear plant remained at about 1.5 cents between 1976 and 1978.

There are a couple of reasons for that. From the beginning, the federal government has paid many of the costs of nuclear power. According to the United States Department of Energy, nuclear energy would cost twice as much—3 cents per kilowatt

The Peaceful Atom

hour—if the government were to withdraw its support. Also, uranium fuel is more convenient, and less expensive, to ship and handle than fossil fuels are because it takes only a relatively small amount of uranium to yield a very large amount of power.

About 6,300 tons (5719.14 metric tons) of uranium will produce 250 billion kilowatt-hours of electricity. To produce the same number of kilowatt-hours, an oil-burning plant would need 425 million barrels of oil. A coal plant would need 120 million tons (108.93 metric tons) of coal, and a natural gas plant, 2.6 trillion cubic feet (74.6 trillion liters) of gas. The cost of transporting such huge amounts of fossil fuels is tremendous—much higher than the cost of shipping 6,300 tons of uranium.

Getting our electricity from fission saves money, say the advocates of nuclear power. Furthermore, it spares our precious oil reserves so they can be used for other purposes. Oil is essential in manufacturing plastics, for example, and it is also an ingredient in many medicines and fertilizers. Refined into gasoline, it powers autos, buses, trucks, and planes.

New Fuel from Old

But what about our uranium reserves? Is uranium, too, a limited resource—one that will eventually become scarce and expensive? Could that make problems for nuclear power in the future?

Yes and no, say nuclear advocates. To understand what they mean, we'll have to take a closer look at the uranium used in nuclear power plants.

As it exists in nature, most uranium is made up of atoms with 92 protons and 146 neutrons. Adding 92 to 146 gives 238, and this most common form of uranium is called uranium-238. Uranium-238—U-238 for short—makes up more than 99 percent of all natural uranium.

Mixed in with the U-238 is a tiny amount of the isotope uranium-235 (92 protons plus 143 neutrons). U-235 is unstable and highly fissionable. It is the essential fuel in nuclear power plants today. One pound of U-235—a hunk the size of a golf ball—could produce as much energy as 1,500 tons (1361.7 metric tons) of coal.

But U-235, comprising as it does less than 1 percent of uranium ore, is scarce. Scientists say that at the present rate of consumption, the world's known reserves of U-235 may be exhausted in only thirty years. Even so, nuclear physicists have little fear that we will run short of nuclear fuel for generating electricity. That's because fission of U-235 doesn't produce just energy; it also produces more fissionable material. As U-235 nuclei fly apart, some of their neutrons strike atoms of U-238. A few U-238 atoms absorb one neutron, thus becoming U-239. The U-239 decays rapidly, forming a new element, plutonium. Plutonium is highly fissionable and extremely radioactive. It is a manmade element; plutonium does not exist in nature.

Scientists call the process by which uranium transmutes into plutonium "breeding." Someday, nuclear advocates believe, we will get our electricity from nuclear reactors that breed plutonium as fast as they consume fuel. Such breeder reactors will simultaneously turn out energy for the present and fuel for the future.

If the breeder works as planned, mankind may have achieved its dream of limitless energy. Electricity will be abundant. It will be "clean," free of the kind of pollution we're accustomed to from fossil fuel plants. There will be no blackouts, no brownouts, no cold homes in the fiercest winter weather. Everyone can enjoy air-conditioned summers. Since oil won't have to be used to produce electricity, it will be available to fill other needs, like fueling automobiles. There will be no lines at the gas station. Factories will hum. There will be jobs for all, and consumer goods will fill our stores. Our way of life will grow better year after year after year. The nuclear dream seems attractive, beckoning us now as it beckoned H. G. Wells in 1913.

Yet there are those who reject the dream. There are those who are doing all they can to force every nuclear power plant in

The first unit of a nuclear power plant in Seabrook, New Hampshire, is scheduled to go on line in 1982. Seabrook has been the site of several antinuclear demonstrations aimed at forcing an end to construction there.

the country to close its doors forever. Some are writing books and articles about nuclear power, or speaking out against it at public meetings. Others are taking part in antinuclear protest marches and demonstrations. A few are trying to halt new plant construction by invading plant property and interfering with the work going on there. To these people, nuclear power is not the answer to the nation's energy problems. It is instead a problem in itself, neither clean nor safe, but the source of a deadly danger, the emitter of the ultimate pollutant—radioactivity.

4 THE WHYS AND HOWS OF RADIATION

Of course, radioactivity doesn't come only from nuclear power plants. Radioactivity has been around since the beginning of time, and human beings have always lived with it.

Natural radioactivity on the earth comes from space, and from the sun. Cosmic rays are produced from radioactive reactions in outer space. Rays of ultraviolet light are produced by radioactive reactions on the sun. Billions of years ago, radioactive waves and particles bombarded the earth continuously. Radioactivity was built into our world as it formed.

Time passed, and things changed. Miles above the earth, a layer of ozone gas began to accumulate. Eventually, the ozone layer thickened enough to keep out most of the ultraviolet light. On earth, radioactive elements were decaying—changing over and over again, through half-lives of hundreds of thousands or millions of years, into the stable, nonradioactive elements we know today.

Yet even now, some of that ancient radioactivity remains. Of the elements that occur in nature, twelve are known to be radioactive. Among them are radium, polonium, and uranium. In addition, there are hundreds of radioactive isotopes of normally stable elements. Scientists have learned to create such isotopes artificially. But under the right conditions, radioactive isotopes are produced in nature as well.

We are surrounded by radioactivity. Our very bodies contain radioactive isotopes of the elements carbon and phosphorus. Construction materials like granite and stone show traces of radioactivity. Soil and water do, too. Even the air we breathe holds small amounts of radioactivity. The atmosphere still permits some radioactive particles to reach the earth, and all of us receive a daily dose of cosmic and ultraviolet radiation. The dose is slightly lower for people living at sea level than for those whose homes are at higher elevations.

Radioactivity from sources like the sun and from natural elements around us is called background radiation. It accounts for about half the amount of radioactivity that the average American absorbs in a year. The other half comes from manmade sources.

One source is television. Color sets especially emit a tiny but steady stream of radiation (which is one good reason for not sitting too close to the TV set while watching). Jet plane travelers get more radiation than the earthbound do. Astronauts get even more. The most significant source of manmade radiation, however, is medical X rays. The gamma ray-like emissions given off during X-raying can be absorbed by the person running the machine, as well as by the patient.

People learned about radioactivity in the 1890s, but it was some time before they realized how dangerous it can be. Many of the first scientists to study radioactivity complained of not feeling well—of having headaches and being tired and depressed—but few recognized the link between their health and their work. Pierre Curie was probably suffering from radiation poisoning when he was killed in an accident in 1906. Nearly thirty years later, his wife, Marie, died of an anemia caused by exposure to radioactivity. The Curies' daughter, Irène, died of the same kind of anemia. Irène Curie and her husband were the first to create a radioactive isotope in the lab.

The early radiation researchers may not have known why so many of them showed similar symptoms, but we do. Radioactivity attacks the cells of the body and changes them. That can lead to disease and death.

The changes are electrical. The atoms that make up our body cells, like all the atoms in all the elements on earth, have electrical properties. Each contains protons and electrons. When an alpha or beta particle, or a gamma ray, strikes a cell, it alters the electrical charge of whatever part of the cell it hits. The alteration is called ionization, and another name for radioactivity is ionizing radiation.

Radiation Safety Standards

By the mid-1930s, scientists were beginning to understand the dangers of ionizing radiation and to urge governments to set limits on the amounts of radiation to which people who worked with radioactivity could be exposed. To express the proposed limits, scientists spoke of curies, roentgen units, rads, and rems. These terms describe the amount of radiation an object gives off, or the amount a person or object is absorbing.

A curie tells us how fast a radioactive material's nuclei are breaking down. That is, it indicates the rate at which radioactivity is being emitted. The higher a material's curie number, the more radioactive it is.

The term "roentgen" comes from the name of Wilhelm Roentgen, the German physicist who discovered X rays in 1895. A roentgen is a measure of how much ionization occurs in a target as radiation strikes it.

Scientists use the abbreviation "rad" to describe how much radioactive energy is being absorbed by a target. "Rad" stands for *r*adiation *a*bsorbed *d*ose. Another abbreviation, "rem," is short for *r*oentgen *e*quivalent in *m*an. It is used to describe the physical effect of radiation on living things. Rems, rads, and roentgen units are very similar, and the terms are sometimes used interchangeably.

Twenty years after scientists first urged radiation exposure standards, United States government limits were publicly announced. Six yars later, in 1960, the Environmental Protection Agency (EPA) set a limit of 3 rems during any three-month span as the maximum to which an individual should be exposed. That amounts to a 12-rem-a-year dose.

For workers in the nuclear power industry, the yearly allowable dose is lower. The Nuclear Regulatory Commission (NRC), the federal agency that makes rules for the commercial nuclear industry, has its own radiation standards. NRC rules limit a worker's yearly dose to 5 rems. For a worker who gets the maximum dose every year over the course of a forty-year career, that means a lifetime dose of 200 rems.

How does such a dose compare with the amount of radiation the rest of us receive in a year? The average American is exposed to about 0.125 rem of background radiation annually. He or she is exposed to another 0.125 rem or so from manmade sources of radiation. Exposures vary, though. They are higher than average for people undergoing a series of dental or medical X rays. Other factors affect radiation dosage. In general, residents of mile-high Denver, Colorado, get bigger doses than people who live on the East Coast. A person whose work place is built of granite—for example, a file clerk in the basement of a large office building—may be exposed to slightly more radioactivity than the person who works on a higher floor, surrounded by wood and plaster.

How do EPA and NRC standards compare with the amounts of radioactivity that are known to damage the human body? A dose of 3,000 rems is almost instantly lethal. A dose one third that size kills within thirty days. Reducing the dose to 450 rems will cause death within a month in about half the people exposed. The other half will become ill, but many will recover. During the three-month recovery period, their general health will be poor, and many may succumb to other diseases. Those who do survive will never regain their full health.

A 250-rem dose will not bring about many immediate deaths. But the exposed population will suffer radiation illness, and a large percentage will die within a few years of radiation-induced cancers.

Low-Level Exposures

Below the 250-rem level, it's hard to predict what radiation effects will be. In very small amounts, ionizing radiation produces changes that a person may be unaware of for many years. Long after exposure, a person may develop leukemia or a cancerous tumor. Or the radiation might damage a gene within an exposed person's body. If that person passes the damaged gene along to a son or daughter, the child may be born with a physical or mental handicap.

Once, scientists believed there was a radiation "threshold," and that exposures below that threshold could not harm anyone. Now, most have changed their minds. They think there is no absolutely safe radiation level, because just one particle of ionizing radiation striking one cell can bring about a deadly change. What's more, a forty-year dose of 5 rems per year appears to have the same biological effect as a single 200-rem dose. Each seems equally likely to lead to disease or genetic damage. But scientists cannot agree on *how much* damage we can expect lowel-level radiation to do to the general public.

One reason why they cannot agree is that they don't have all the facts they need. Commercial nuclear power is only about twenty-five years old, and in 1980, nearly three fourths of this country's nuclear plants had been in operation less than ten years. It takes thirty or forty years for many radiation-induced cancers to appear. Similarly, it may take more than one generation for certain genetic effects to become apparent. There simply hasn't been enough time to collect all the data needed to determine what the hazards of manmade radiation really are.

Another reason for the scientific disagreement is that it is impossible to prove that a human cancer or genetic alteration is due to this or that particular source. Radiation from nuclear power is only one possible cause of cancer and birth defects. Other suspected causes are numerous: background radiation, certain chemical wastes, food additives and preservatives, cosmetics, some drugs and medicines, and so forth. No one knows for sure which cause is responsible for each illness or defect among the public. Tumors and abnormalities don't come with handy labels that tell whether or not they were caused by radiation from commercial nuclear power.

So the question of how dangerous low-level radiation is has no definite answer. It may never have one. But people don't need a final answer to know that radioactivity can cause disease and death, and that there is no radiation threshold of safety. Nuclear power plants are designed with this knowledge in mind. They are built to keep radioactivity away from the public.

Containment—How Workable?

Keeping radioactivity contained within the plant means isolating the reactor core—the source of radioactivity—from the outside world. Uranium is only one of the many radioactive elements present in the core. Plutonium, bred when neutrons hit U-238 nuclei, is there, too. Nuclear reactions create fission products as well. Among them are krypton-85, strontium-90, iodine-131, cesium-137, and iron-59.

From the fuel rods in the core, radioactivity spreads. The water or steam that acts as moderator and coolant becomes radioactive. So do the pipes that carry the water, and the huge steel vessel in which the steam is heated. In fact, everything that comes in contact with the core's radioactivity becomes radioactive itself.

To keep radioactivity from getting out and harming the public, the reactor core is housed within a 9-inch(22.85 cm)-thick steel vessel. This is surrounded by a concrete shield. The reactor and its housing, along with the steam-pressure vessel and the other radioactive parts of the plant, are sheltered within a huge containment structure. A typical containment structure is 300 feet (91.44 meters) high and 130 feet (39.62 meters) in diameter. A steel shell lines its inner wall. Its domed outer wall, the one you see as you pass a nuclear plant, is of reinforced concrete, 3 to 5 feet (.9 to 1.5 meters) thick.

Entrances and exits in the containment structure are locks and hatches especially designed to keep radioactivity from leaking out. Where it is necessary for steam pipes or other fittings to penetrate the walls, openings are equipped with elaborate seals and valves that are supposed to keep toxic liquids and gases within the dome.

Yet even with all these precautions, some radioactivity escapes. Industry officials acknowledge that radioactive containment can never be 100 percent effective. So Nuclear Regulatory Commission regulations allow nuclear power plants to release some radioactive gases into the atmosphere. They also permit mildly radioactive water to be pumped back into whatever natural body of water it came from.

NRC regulations do require that each radioactive discharge be "as low as practicable"(ALAP). According to spokesper-

The Whys and Hows of Radiation

sons for the industry, ALAP standards ensure that routine emissions will expose a person standing just outside a plant to no more than 0.005 rem a year of radioactivity. Overall, they say, routine emissions from nuclear plants contribute no more than 0.00001 rem a year to the average American's exposure.

However, routine emissions are not the only radioactive releases from a nuclear power plant. Unplanned emissions of radioactive gases and liquids also occur from time to time. They are the results of problems within a plant.

What sort of problems? A gasket might blow, flooding the reactor room with thousands of gallons of radioactive water. If the water is not too radioactive, it may be sent back into the lake, river, or bay from which it came. If it is more radioactive, workers may decontaminate it—remove most of the radioactivity from it by chemical means—before pumping it back. Other problems might be defective seals, leaky pipes, faulty wiring, missing or broken parts, or human error. All are known to have led to emissions of radiation from even the best-planned containment structures.

Each inadvertent radioactive release may be very small indeed—on the order of 0.00015 rem at a time, perhaps. If plant officials tell news reporters about an accidental release, they will dismiss it as "tiny," "minor," or "insignificant."

Is it really? In one sense, yes. By itself, a 0.00015-rem dose is an insignificant fraction of the 0.250 rem or more of ionizing radiation that most Americans receive in a year.

But that isn't the whole story. An accidental 0.00015-rem dose of ionizing radiation cannot be considered all by itself. There may be many accidental releases from a single plant each year. These must be added to our background dosage, and to whatever amount of radioactivity a plant releases in the normal course of events. And it's not a matter of emission from only one plant. As the United States entered the 1980s, 70 nuclear plants were operating within its borders. Ninety more plants were on order or under construction. There were more than 220 plants worldwide. That number, too, is expected to grow, and as it does, more and more plants will be emitting radioactivity, both routine emissions and accidental releases. Much of that radioactivity will be around for a long, long time.

How long, depends on the half-life of each radioactive isotope. Krypton-85, released as a gas from a nuclear plant, has a half-life of 10 years. Let into the atmosphere, it will be reduced by only 50 percent at the end of a decade. Scientists tell us that a radioactive substance is no longer dangerous after 10 or 20 half-lives. So krypton-85 accidentally released in 1985 will become harmless sometime in the year 2085 or 2185. Unfortunately, most of us won't be around on that happy day. The krypton will remain radioactive throughout our lifetimes.

Other commonly emitted radioactive substances disappear more quickly. Iodine-131's half-life is just 8 days. Iron-59's is 45 days. Still other isotopes stay in the environment far longer. The half-life of strontium-90 is 28 years. Of carbon-14, 5,770 years. Of nickel-59, 80,000 years.

Radiation and the Environment

What happens as released radioactivity ticks its half-lives away is determined by chemistry. Iodine-131, for example, does just what stable iodine does. Released into ocean waters, it concentrates in seaweed. Tiny sea animals, such as shrimp, feed on the seaweed, and the radioactive iodine enters their bodies. Small fish eat the shrimp—and the iodine. In their turn, the small fish are eaten by larger ones. This progression, plants being eaten by animals, and smaller animals being eaten by larger animals, is known as the food chain.

Steadily, the iodine-131 progresses up the food chain. It may end up in a can of pet food or in a sack of fertilizer. It may make its way onto someone's dinner plate. When that person eats it, it will do what iodine always does in the human body. It will migrate to the thyroid gland and stay there. Eventually, the person could develop cancer of the thyroid.

Other radioactive isotopes follow equally predictable routes. Strontium-90, which chemically resembles calcium, gets into milk. A dairy farmer who lives near a nuclear plant doesn't have to wait to read in the papers that strontium-90 has been accidentally emitted in his neighborhood. He can tell by checking the strontium level of his herd's milk. When people drink the milk, they accumulate strontium-90 in their teeth and bones.

Radioactive iron-59 acts exactly like the nonradioactive iron that is essential to human health. It goes into the body's red blood cells and travels throughout the circulatory system. Radioactive cesium-137, which has chemical properties like those of the element potassium, becomes, like potassium, part of our muscle tissue. Inevitably, as certain radioactive chemicals are released into the environment, they make their way into the food chain, with results that no one can be sure of until we know more about the effects of low-level radiation on human health.

Radioactive elements that don't get into the food chain can alter the environment—and threaten life—in other ways. Krypton gas, for instance, is chemically inert (inactive), so it does not enter the food chain. Instead, it accumulates in the atmosphere, where, scientists fear, it may produce hard-to-predict electrical changes. The changes could affect the formation of raindrops, which could mean drastic modification of weather and climate patterns. Krypton was one of the radioactive gases released during the cleanup after the March 1979 accident at the Three Mile Island nuclear plant in Pennsylvania. Plant officials assured the public that the krypton would not harm them, but many of the nation's leading scientists are not so sure. No one knows what the effects of radioactive krypton may be, they say.

Questions of Safety

But if the public is threatened by radioactivity from nuclear plants, what about the people who work inside the plants? For them, the danger must be greater and more immediate. No massive dome of steel and concrete stands between them and the fission products at the reactor core. These men and women spend their working lives in a space where radioactivity spreads and becomes more intense with every year that the reactor is in operation. So the equipment the workers use and the procedures they follow are designed to keep their exposures within the 5-rem-per-year limit set by the NRC.

One precaution is to have workers do many jobs at a safe distance from the core. Using remote-control devices, technicians can change steam pressure, alter water levels, regulate the rate of the chain reaction—even bring the reaction to a standstill—without absorbing excess radiation.

Another precaution is to stop operating a plant when workers must do certain jobs. One job that requires a complete shutdown is replacement of used fuel rods. Even after three or four years in the core, some U-235 is left in the fuel rods. But by then, the rods also contain accumulated fission products that interfere with the chain reaction, slowing it down so it gives off less and less energy. Workers must remove the spent rods and put new ones in their place. Refueling takes a month or two each year, and while it is going on, the plant produces no electricity.

Other jobs aren't complex enough to justify a shutdown, nor can they be done by remote control. Most are routine maintenance jobs like fixing broken pipes or repairing worn-out valves, and they require workers to enter the plant's "hot"—radioactive—areas. These workers must wear protective coverings. The kind of covering they wear depends on the degree of radiation to which they are exposed.

Replacement of spent fuel rods is one job that requires a complete plant shutdown. Here, a huge mechanical "arm" puts a new fuel assembly into the core.

Before entering such areas, workers must don paper or cloth lab coats and caps, paper or rubber boots, and gloves. They carry radiation-sensitive badges that allow them to keep track of how much radioactivity they are absorbing. When the job is done, the workers leave the plant's hot area, strip off the protective clothing, and toss it into barrels to be laundered or thrown away. Then they pass through radiation detectors to learn whether or not any parts of their bodies have become contaminated with radioactivity. If they have, they scrub the contaminated area with soap and water.

For jobs in "hotter" areas, the protection of paper or cloth isn't enough. Workers are encased in rubber clothing, similar to a diver's wet suit, and they may have to wear gas masks.

Even that isn't sufficient protection for spots where radiation levels may be 25 or more rems an hour. At such levels, it would take only a few moments for a worker to absorb the entire annual radiation dose allowable under NRC rules. Yet the job may require several hours to complete.

One solution is to use teams of workers. The first worker puts on protective clothing, goes into the hot spot, and starts the job. A minute or two later, he's replaced by a second worker, then by a third. Working this way, it can take scores of people to perform even a simple task. At one Illinois plant, a job that could ordinarily have been done by a dozen men required 350 men—each working only a few seconds—to complete. At a New York plant, it took 1,700 welders a full six months to repair a single broken pipe.

Careful Enough?

Even with all the elaborate precautions, many in the nuclear power industry fear that their workers may be getting too much radioactivity. So they rely more and more on "jumpers" to take some of the dose.

Jumpers are men and women hired on a one-time basis to do

Worker safety is a concern at nuclear plants. This man wears protective clothing to do jobs on the "hot" side of the plant. He places the equipment he's finished using in a plastic bag for decontamination or disposal.

just a few moments' work under highly radioactive conditions. Jumpers can earn a lot for a job—up to $75 for work that may take only a few minutes to perform. They absorb a lot of radiation, too, usually 1 rem or more at a time. Many jumpers go from plant to plant, doing a job here and a job there, picking up an unknown amount of radiation in the process.

According to NRC estimates, jumpers are getting at least half the total radiation dose of the country's nuclear plants. More precise figures are hard to come by. Industry reports on jumpers, on how many are employed and what dose each receives, take about three years to be made public. That makes it difficult to judge whether or not individual jumpers are getting more than the 5-rem-per-year limit. One thing is clear, however. The use of jumpers doubled between 1973 and 1976, and it is likely to continue to rise as the nation's nuclear plants grow older and their radiation levels increase.

Many people criticize the industry's reliance on jumpers. They point out that using jumpers simply spreads the radiation dose out over more and more people. Giving a 5-rem dose to a hundred people is as undesirable as giving a 20-rem dose to twenty-five people. Instead of hiring jumpers, critics say, industry officials ought to concentrate on making the plants safer and less radioactive for *all* workers.

That's not the only criticism leveled at plant officials' treatment of their workers. Accounts of routine carelessness at nuclear plants are common. Some have been recorded by organizations like the Massachusetts Public Interest Research Group (MPIRG).

The MPIRG alleges that at a plant in Nebraska, workers were needlessly exposed when the door to a room in which radiation was 5 or 6 rems an hour was propped open and left that way. In Oregon, two technicians entered an area in which radiation levels had not been checked. MPIRG reports that one man got 17 rems; the other got 27. In a Wisconsin plant, a supervisor spent thirty seconds in a radiation field later measured at 2,000 rems an hour.

Another Massachusetts-based group, the Union of Concerned Scientists (UCS), tells of a plant in which drinking water became contaminated after a hose was connected the wrong way. The UCS has documented another case in which a basketball was used to seal a reactor pipe. The ball worked loose, allowing 14,000 gallons (52,990 liters) of radioactive water to gush out.

In a way, it's easy to understand why such exposures occur. Nuclear radiation gives no warning of its deadliness; plant workers cannot see it, hear it, taste it, feel it, or smell it. A person can receive a lethal dose without sensing the slightest hint of danger.

Nuclear industry representatives defend themselves, and the safety of their plants, vigorously. They point out that no civilian worker has ever died of acute radiation poisoning in a commercial nuclear power plant in the United States. Three men did die in a reactor accident at the Idaho National Engineering Laboratory in 1961. But that was a test reactor, not a commercial one. And other physicists have been killed by radiation while working in America's nuclear weapons program. However, most of America's 50,000 nuclear workers feel perfectly safe. A small number have resigned because of what they feel are hazardous conditions, but the vast majority are comfortable in their jobs. They—and their employers—believe that plants are safe now, and that they will become even safer as the industry develops more sophisticated radiation detection and monitoring systems.

That may be true. But even if it is, it will solve only part of the safety problem of commercial nuclear power. For the industry faces safety concerns that have nothing to do with power plants and their reactors. These concerns relate to other parts of the nuclear "fuel cycle."

5
THE BROKEN CYCLE

The nuclear fuel cycle has two parts. At the "front end" of the cycle is everything that happens to uranium before it reaches the reactor. The "back end" includes what becomes of spent fuel rods and the radioactive wastes that each one produces.

The front end of the cycle begins with uranium mining. The world's richest sources of uranium ore are in South Africa, Australia, the Soviet Union, Canada, and the United States. American uranium mines are located in the western states, mostly in New Mexico, Wyoming, Colorado, Utah, and Washington. Uranium ore has recently been discovered in the northeast as well.

Uranium ore is naturally radioactive. It contains the element radium, which emits alpha particles. Breathed in by those who dig it out of the ground, radium can cause lung cancer or other respiratory illnesses. Uranium also contains radioactive radon, a gas given off as uranium undergoes radioactive decay. Radon, too, causes lung disease.

Radon and radium have claimed many victims among American uranium miners. A federal government study of thirty-five hundred men who mined the ore during the 1940s and '50s shows that two hundred of them had died of lung cancer by 1979. That's five times the lung cancer rate of the rest of the population. The rate is expected to climb higher as more miners develop cancer symptoms. Declares a public health doctor in New Mexico, "There is an epidemic of lung cancer among former uranium miners."

The health picture for uranium miners should be brighter in the future. New laws are forcing mining companies to keep shafts and tunnels freer of dust and to monitor workers' exposures more closely. And today's miners are more likely than earlier ones to be aware of radiation hazards and to try to avoid needless contamination.

Milling and Mill Tailings

Milling is the next step in the fuel cycle. In milling, concentrated uranium is separated from uranium ore. This is accomplished by crushing the ore, grinding it to a powder, and treating it with chemicals. The result is "yellowcake," uranium concentrate in solid form.

Another result of milling is mill tailings—the ore material that remains after the uranium concentrate has been removed. The tailings contain most of the radioactivity of the original ore. This radioactivity is present in radium and radon, in thorium (which has a half-life of 80,000 years), and in other decay products. The men and women who work in uranium mills encounter health hazards similar to those faced by uranium miners.

The level of radioactivity in the tailings is low. But the sheer

A uranium milling plant in Gas Hills, Wyoming

quantity of the tailings is high. The NRC says that the nation's twenty-one uranium mills have accumulated over 140 million tons (127.09 million metric tons) of tailings. That figure will rise to about 750 million (680.85 million metric tons) by the year 2000. Nearly all tailings are in the western United States, since uranium mills are built next to uranium mines.

Until recently, not much was done with the tailings. Generally, they have been left in huge piles around the mills. When the wind blows over them, it picks up radioactive dirt and dust and carries it for miles across the countryside. When it rains, radioactivity is washed into streams and rivers.

Thus scattered, the radioactivity works its way up the food chain. It concentrates, according to its chemical properties, in plants, fish, wildlife, or livestock. Studies of plant and animal life along rivers contaminated by uranium tailings show above-nor-

At a plant in New Mexico, yellowcake is prepared for shipment.

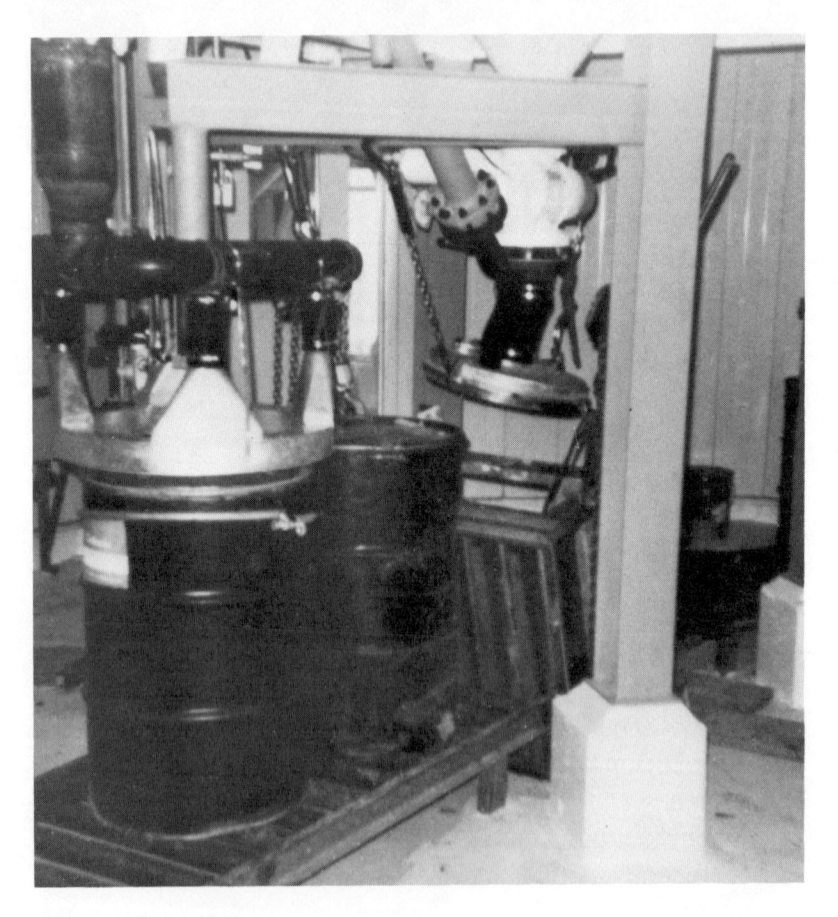

mal levels of radiation as far as fifty miles (80.45 km) downstream from mill sites. Radioactivity seeps into people's wells, making the water unfit to drink.

Not all tailings have been left in heaps to be blown and washed into the environment, however. During the 1960s, city officials in Grand Junction, Colorado, thought of a way to get rid of the huge piles near one abandoned mill. They utilized the tailings as landfill and added them to concrete.

The concrete was used to construct various buildings, including homes and schools. By 1970, an increase in birth defects among children born to parents living in houses whose foundations contained uranium tailings caused health officers to wonder about radiation levels in the buildings. A check revealed that the levels were high.

In 1978, Congress passed the Uranium Mill Tailings Radiation Control Act, which gives the NRC direct control over tailings at operating mills and allows the NRC to cancel the license of any mill that does not safely dispose of radioactive tailings. The law also gives the NRC the power to advise the federal government's Department of Energy on how best to clean up tailings piles at mills that are no longer in operation.

How effective the Tailings Control Act will prove to be remains to be seen. A year after it went into effect, a mill in New Mexico accidentally sent 1,100 tons (998.58 metric tons) of uranium tailings and 100 million gallons (378.5 million liters) of radioactive water into a small brook known as the Rio Puerco. The brook flooded, and as the waters raced downstream, the curie count in the Puerco soared to almost 7,000 times the limit accepted as safe. Dangerous levels of radioactivity were found in wells within 200 feet (60.96 meters) of the brook. Public health officials worry that this radioactivity may get into local ground water, poisoning wells at greater and greater distances from the mill.

This spill occurred at a fully licensed, almost new plant that was considered exceptionally safe and well designed. Yet, not

A pile of uranium tailings in Colorado. Recent legislation gives the NRC control over tailings. Will that make the front end of the nuclear fuel cycle safer?

even the newness and good design were enough to prevent a serious accident at this point in the front end of the nuclear fuel cycle.

Gaseous Conversion, Enrichment, and Fabrication

Other dangers crop up as the nuclear fuel cycle continues. In the next step, the yellowcake is converted into a gas. There are two gas-conversion plants in the United States, one in Illinois and the other in Oklahoma.

From Oklahoma or Illinois, gaseous uranium goes to one of the nation's three enrichment plants. These plants, located in Tennessee, Kentucky, and Ohio, are owned by the federal government and operated by private companies.

At the enrichment plants, the uranium gas is treated to increase its concentration of fissionable U-235. The gas is forced through a series of very fine filters. U-235 atoms pass easily through the filters, but the U-238 atoms, being larger and heavier by 3 neutrons, have a harder time. Some are trapped, lowering the number of U-238 atoms in the gas. Eventually, the gas is about 97 percent U-238 and 3 percent U-235, a concentration of U-235 that will be high enough to sustain a chain reaction when the time comes.

Enrichment produces still more radioactive tailings. For every 5 pounds (2.268 kg) of yellowcake that enter an enrichment plant, only 1 pound (.4536 kg) emerges as usable, U-235-rich fuel. The other 4 pounds of depleted uranium are just another waste product.

The final step at the front end of the fuel cycle is fuel fabrication. In this step, uranium is reconverted into a solid, made into pellets, and packed into the zircalloy rods. The rods are shipped to the country's nuclear power plants to await their turn to be placed in position in the reactor cores. The front end of the cycle is complete.

Back End of the Nuclear Fuel Cycle

The back end of the nuclear fuel cycle begins when spent rods are removed from the reactor core. The spent rods are intensely radioactive. Of the radiation produced during each one's lifetime, 99.99 percent remains in the rod. Penetrating beta or gamma radiation is given off by most of the hundreds of fission

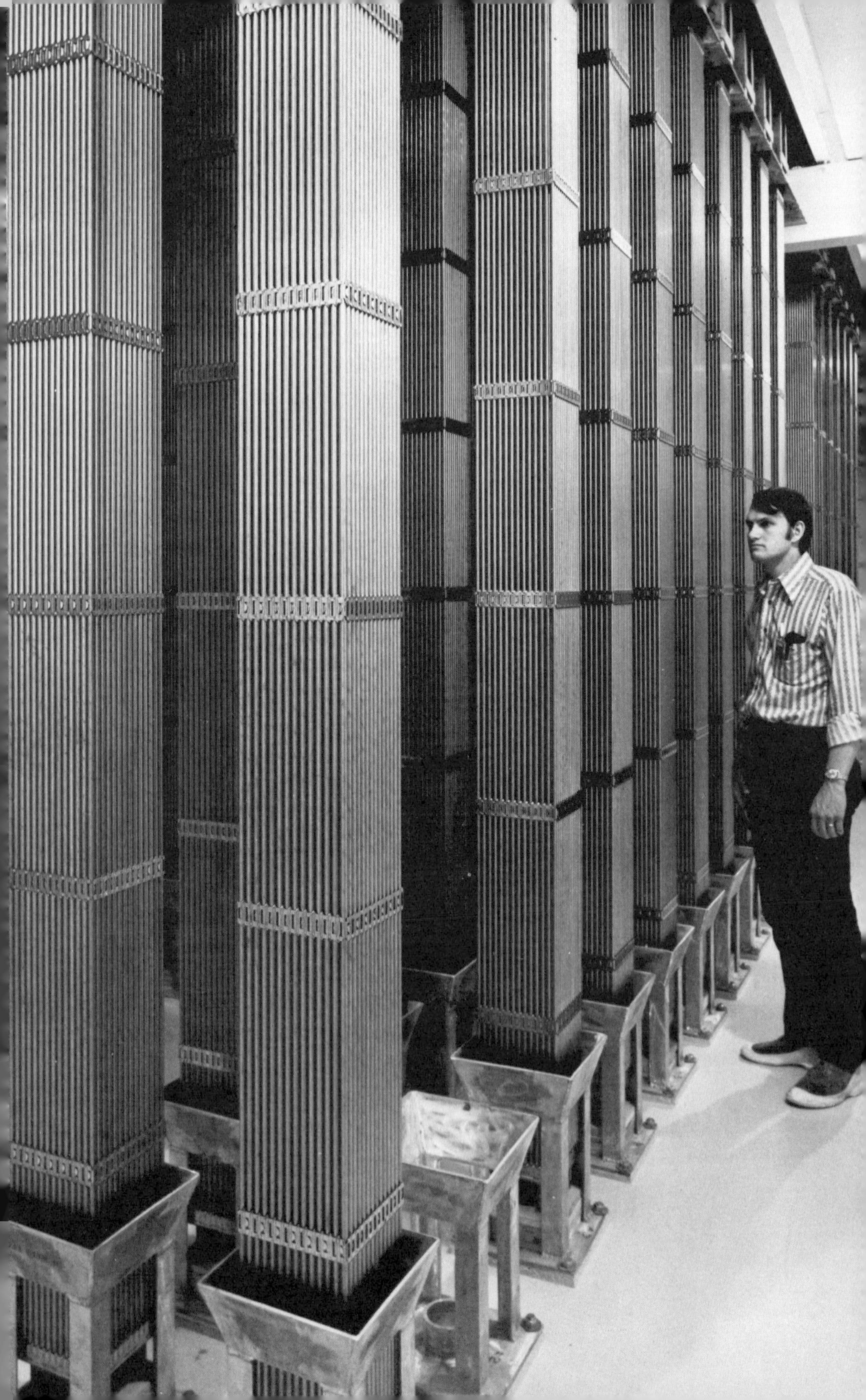

products in each rod. Plutonium, also present in the rods, emits alpha particles.

Not only are the rods radioactively hot, they are hot in temperature. To cool them, workers store the rods underwater in large pools next to the reactor room. The pools, ruggedly constructed of steel and concrete, and flooded to a depth of 30 feet (9.144 meters) with a boric acid solution, are designed to keep radiation from escaping into other parts of the plant. The water in a storage pool glows with an eerie blue light, not a light caused by radiation (radiation cannot be seen) but by the immensely high speed at which electrons are moving.

After three to six months in the pool, the rods have lost a great deal of their heat. They have lost some of their radioactivity, too, as the isotopes with brief half-lives become harmless. Now the rods are ready for the next step: reprocessing, which is the recovering and recycling of used fuel.

That is, they would be ready for reprocessing if any reprocessing were going on. For nearly a decade, reprocessing has been a missing step in the commercial nuclear fuel cycle in this country.

Reprocessing, the Missing Step

Prior to 1972, reprocessing took place at a plant in West Valley, New York, not far from Buffalo. The West Valley plant, which opened in 1966, was the only commercial reprocessing facility ever to operate in the United States.

At West Valley, the 12-foot (3.6576-meter) rods were cut into shorter lengths and immersed in a nitric acid "bath." The acid dissolved the fuel inside the rods, allowing workers to

At every nuclear plant, new fuel rods wait to be placed in the reactor core. Packed with half-inch-long uranium pellets, each 12- to 14-foot rod will produce energy for three or four years. Then it will be replaced with a new rod.

The back end of the nuclear fuel cycle begins when spent rods are placed in temporary storage in large steel and concrete pools. But "temporary"

s turning into long-term; there are neither operating reprocessing plants nor permanent storage facilities in the United States today.

The Broken Cycle 57

recover the materials within each one. Among the matter recovered was unused, but still fissionable, U-235 and the plutonium produced during the chain reaction. The uranium and plutonium could be isolated from the other fission products, refashioned into new fuel pellets, and used to generate more electricity.

Reprocessing—and reusing—spent fuel is a vital part of the nuclear fuel cycle. As we saw, the world supply of fissionable uranium might be used up in about thirty years. Unless the supply can be stretched, or new supplies discovered, nuclear power has just a thirty-year future.

But commercial reprocessing has not been a success in the United States. From the start, the West Valley plant was in trouble with federal regulators. According to them, plant workers were poorly trained and contamination was a constant problem. After repeated shutdowns because of safety violations, the plant halted operations entirely in 1972. At first, the owners of the plant insisted they were planning to enlarge and reopen the facility. But in 1976, they announced that the closing would be permanent.

The next year, reprocessing suffered another blow. President Jimmy Carter announced that he opposed setting up any new reprocessing plants. Carter expressed the fear that plutonium extracted during reprocessing might be used by other countries to build nuclear weapons.

The President's motive for opposing reprocessing—to prevent the proliferation of nuclear arms—may have been good, but there are those who quarrel with his action. Physicist Robert O. Pohl, of Cornell University, argues that reprocessing is essential because it gives the nuclear industry a safe way to use up much of the dangerous plutonium created during fission. He calls reprocessing "our moral obligation toward future generations."

Dr. Pohl has a point. Every operating nuclear power plant produces from 400 to 500 pounds (181.44 to 226.8 kg) of plutonium a year. For the country's seventy plants, that means anywhere from 28,000 to 35,000 pounds (1270.08 to 1587.60 kg) annually. Compare those numbers with the medical estimate that less than 1/1,000,000 of a gram of plutonium is enough to cause lung cancer. And the plutonium produced in our nuclear power plants will be around for a long time. This manmade element has a half-life of 24,400 years.

The President had a point, too. Plutonium collected from a reprocessing plant could be used in a bomb. Aside from that danger, a reprocessing plant like the one in West Valley poses the threat of contamination to its workers and the public alike. A further problem is that although reprocessing may help to use up some of the plutonium produced in nuclear power plants, it does not get rid of it all. Nor has reprocessing any effect on the many other deadly isotopes in spent fuel rods. Reprocessing itself produces enormous amounts of extremely dangerous wastes, including the inert gas krypton. For every pound (.4536 kg) of usable plutonium extracted in reprocessing, there will be 200 pounds (90.72 kg) of waste material. Eight years after the West Valley plant closed, state and federal officials were still wondering what to do with the 560,000 gallons (2,119,600 liters) of highly radioactive liquids and the 170 tons (154.326 metric tons) of spent fuel rods at the site. Should the wastes be allowed to stay there, a threat to local residents? If not, who should pay for cleaning them up and transporting them to a storage area? That's a problem, too; scientists estimate the West Valley cleanup will cost from $180 million to $1.1 billion.

Waste Storage

The fact that reprocessing is a missing step in the nuclear cycle means that spent fuel rods have nowhere to go. Instead of lying in their steel-and-concrete pools for a few months, rods are accumulating there year after year. At some plants, storage pools have been made bigger. At others, federal regulators are permitting the space-saving technique of placing the rods closer and closer to each other. That practice could be dangerous. If enough fissionable U-235 remains in the spent rods, and the rods are packed tightly enough together, they could act like a bundle of fuel rods in the reactor. Neutron-absorbing material must be placed between the rods to prevent a chain reaction. Even with this precaution, however, some people fear that the rods could spontaneously go critical.

Under the circumstances, it's not surprising that the nuclear industry is crying out for help. Industry leaders know that by 1980 there were at least 2,535 tons (2301.27 metric tons) of spent fuel rods in temporary storage around the country. Unless something is done, there will be 108,000 tons (98,042.4 metric tons) by the year 2000.

Industry officials are eager for the federal government to allow reprocessing to begin again. Equally, they want the government to start preparing new dumping grounds for high-level wastes. Unless new dumping grounds are found, the nation's highly profitable nuclear industry cannot survive for long. Today, just three dumps serve the needs of all the country's nuclear power plants. The dumps also receive low-level wastes from medical schools and hospitals.

One dump is in Barnwell, South Carolina. Ninety percent of the nuclear industry's low-level wastes are trucked to the Barnwell site. These wastes include mildly radioactive liquids and gases, contaminated plant parts, and protective clothing and equipment used by plant employees. Each plant produces up to 14,000 cubic feet (392 cubic meters) of low-level waste in a year. A second dump is located in Beatty, Nevada, and a third in Hanford, Washington. Hanford is the only commercial dump licensed to receive high-level wastes like plutonium.

Nuclear Dumps Today

Conditions at the nation's three nuclear dumps have not been good. In some cases, radioactive items were buried in trenches so shallow that the trash has been uncovered by wind and water. People who live nearby, not realizing that the exposed articles are radioactive, have helped themselves to tools and clothing. In other cases, notably at Hanford, liquid wastes have seeped from damaged or corroded containers.

The leaking radioactivity gets into the environment, just as surely as does radiation released at earlier points in the fuel cycle. In Hanford, for instance, it was found that jackrabbits were dropping radioactive pellets. An investigation showed that the rabbits became contaminated after burrowing into the ground near leaking containers. Before long, radioactivity was appearing in the coyotes and hawks that feed upon these jackrabbits.

In 1979, the states of Washington and Nevada acted to protect their citizens from radioactive poisoning. The governors of the two states temporarily closed their dumps to shipments of nuclear wastes from power plants. Washington's governor, Dixy Lee Ray, was particularly distressed that wastes were not even arriving at Hanford in safe condition. Governor Ray was a former chairwoman of the Atomic Energy Commission, forerunner of the NRC, and had long been a promoter of nuclear power. But she was appalled by the conditions under which radioactive materials were being transported around the country. She cited instances of containers that spilled their contents over the floors of the trucks that carried them, and radioactive gravel in damaged boxes. Unsafe shipping of nuclear wastes endangers plant and animal life in every town and city on the transport route as well as the people who live near, or work at, the dumps themselves.

The Nevada and Washington dumps reopened shortly after their 1979 closings, but the radwaste problem is far from solved. The Hanford and Barnwell dumps are cutting back drastically on the amount of out-of-state nuclear wastes they accept. Wastes have been collecting around the country for twenty-five years now, and no one knows just what to do with them.

Storage Possibilities, Storage Problems

Suggestions have been made. One is to solidify the wastes in the form of glass or ceramic, and to bury them deep underground. Mine shafts or ancient salt deposits are mentioned as possible burial places. Another suggestion is to excavate the seabed and place some of the wastes below the ocean floor. Some think it might be possible to find a way to transmute toxic substances into stable elements, but so far no one has found a practical method. One Australian geochemist thinks radwastes could be made into elements with rocklike properties, then melted together with real rocks and safely buried. He calls his idea the synroc (synthetic rock) solution.

Storing the wastes beneath the Antarctic ice cap, leaving them on sparsely populated Pacific islands, and shooting them into space are other, less realistic, proposals.

Of the ideas advanced so far, the most promising seems to be the first: vitrifying the wastes, turning them into glasslike materials. Formed into blocks about 6 feet by 18 inches, the wastes would be fitted into canisters of stainless steel. Scientists estimate that the high-level waste from an operating reactor would amount to between five and ten blocks a year. The final step would be placing the sealed canisters underground, perhaps in salt formations that are millions of years old.

Salt deposits are a commonly suggested feature of this proposal. Why? First, because they have remained unchanged for so long. Many scientists assume that the deposits will continue to be geologically stable for hundreds of thousands of years more. Second, because the salt in the deposits has been dry for such a long time, which would indicate that it is totally separated from any ground water. That means no water will seep into and corrode the steel containers.

This waste storage plan has its detractors, however, among both scientists and nonscientists. Members of the Union of Concerned Scientists claim that *no* storage method is safe for ele-

Many Americans, even some who support nuclear power, maintain that radioactive materials are being shipped around the country in inadequate containers. In a laboratory test, this drum, intended for transporting plutonium by air, was slammed against an unyielding target at a speed of 300 miles an hour. Now it must be tested for leaks.

ments like plutonium. The UCS points out that in 1971, the federal government was all set to start storing radwastes in a salt deposit in Kansas, when local geologists learned that salt miners working a half-mile (.80 km) from the proposed waste site had pumped thousands of gallons of water into the ground.

(One way of mining salt is to dissolve it in water and bring it to the surface in liquid form.) But at the Kansas mine, the water did not reappear and about 175,000 gallons (662,375 liters) of it were "lost" below ground. If that water got into the waste site, geologists warned, it might damage the containers and let radioactivity escape. Embarrassed, government officials gave up their plan to bury radwastes in Kansas. No other suitable site has been found, and the wastes are still awaiting permanent storage.

UCS scientists doubt that any really safe storage sites exist. Who can be sure that just because a geological formation has been stable for thousands of years, it will continue so for the enormous length of time it will take for radioactive wastes to decay through 10 or 20 half-lives? Who can accurately predict where floods, earthquakes, and other natural disasters will strike during the next millennia?

Another objection to burial in salt beds, the UCS says, is that salt deposits are often accompanied by deposits of other valuable minerals, including oil and natural gas. Drilling for oil and gas in the vicinity of highly radioactive wastes could disturb those wastes and release deadly radiation. Can the nuclear industry of today guarantee that no one will mine near a radwaste dump over the next half-million or so years?

Disposing of Reactors

Still another radwaste problem haunts the nuclear industry. It is the question of what to do with worn-out reactors.

The nuclear reactors we now get power from will not have long lives. After thirty or forty years of use, their parts will be old and outdated. Fixing or replacing them will be impossible because the years of operation will have made the reactor components too radioactive for humans to work near them. Even if the plants were in reasonable working order, radiation levels under their domes would be so high that the plants would have to be abandoned.

So the nuclear industry, which has yet to deal satisfactorily with its day-by-day wastes, is going to have to dispose of large metal components and concrete structures with 5-foot (1.5-meter)-thick walls. And it will have to do so soon. By the 1990s, the country's oldest nuclear plants will be wearing out.

So far, the industry has considered three possibilities for decommissioning nuclear plants. The first, mothballing, would involve removing radioactive fuel and liquids from a plant. Then the plant would be locked and kept under strict guard until it was no longer radioactive. Estimates vary as to the length of time a guard would be needed. Industry people say two hundred years. But an analysis done for the New York Public Interest Research Group contends that it would take 1.5 million years for plant radioactivity to be rendered harmless.

The second decommissioning possibility is entombment. Concrete would be poured around a plant's radioactive components. Then the plant would be sealed and guarded. It would be easier and less expensive to guard an entombed plant than one that had been simply mothballed. The guard probably wouldn't have to be maintained so long, either.

Dismantling, the third possible solution, would be the safest course. The reactor and its containment structure would be torn down. The area where it had stood would be decontaminated, if necessary, and restored as nearly as possible to its original condition. No guard would be necessary. Of course, the dismantled reactor and parts of its containment structure would have to be safely stored, so this is an unworkable solution unless the nuclear industry finds a way of handling its other waste products.

Will it? That's a key question for the industry. But it's only one of the problems that industry officials face.

6
SAFETY SYSTEMS AND AN ACCIDENTAL TEST

The light bulbs have burned out inside the reactor building. It is dark. The temperature at the core is down from the approximately 5,000 degrees Fahrenheit (2,760 degrees Celsius) of a few months ago, but the reactor room is still warm. Highly radioactive steam rises from an 8-foot-deep pool of spilled coolant. The deadly mist fills the air. Reverberating from wall to wall, now as it has been for months, is the shrill clamor of an alarm bell.

This was the ghostly scene at the Number 2 reactor of the Three Mile Island nuclear power plant in Middletown, Pennsylvania, a year after what is considered to be the country's worst nuclear accident. The accident at the almost new reactor began on March 28, 1979, with a simple equipment failure. It ended weeks later with the reactor at a cold shutdown. That accident will play a large part in determining the future of nuclear power in the United States.

Nuclear critics maintain that the fact that the accident happened at all shows it is too dangerous to go on producing electricity from fission. Advocates of nuclear power contend just the opposite. The events at Three Mile Island, they say, demonstrate how safe nuclear power is. Even when things went wrong, the plant's systems worked to protect workers and the public from hazardous radiation doses.

The safety precautions at TMI, and all nuclear plants, are designed to do two jobs at once. They are intended to keep accidents from taking place in the first place. And if an accident should occur, they are supposed to prevent large quantities of radioactivity from getting outside the plant.

One obvious safeguard we've already seen is that of locating the control room, from which technicians govern the chain reaction, at some distance from the reactor itself. The technicians rely on mechanical devices to raise and lower the control rods, to alter the level of coolant, and to do the other jobs that keep the reaction going smoothly. That way, no technician should be in danger, whatever happens in the reactor, and however hot the core bcomes.

Another precaution is to provide a backup system to take over the instant a plant's human operators make a mistake. Suppose an absent-minded technician overlooks a developing problem. He might fail to see that a gauge indicates a sudden drop in the level of coolant, or to notice that the core temperature is rising precipitously. Automatic computerized equipment doesn't have such memory lapses. As soon as it senses the problem, it will drop the boron control rods into the core, causing the reactor to "trip," or "scram."

Some newer nuclear plants have the additional safety feature of auxiliary controls outside the main control room. There, human control-room operators can make a mistake or the auto-

Safety Systems and an Accidental Test **67**

An amazing array of switches confronts nuclear plant control-room workers. From here, technicians can monitor what is happening at the core. The large circle to the right of the standing man's head shows the position of the control rods, indicating which ones are raised and which are in position in the core.

matic backup system can break down, but workers in another part of the plant can still step in to stop the reaction the moment they sense a problem.

But suppose these workers, too, fail to act in an emergency. Or suppose they act hastily, and in their haste they make a mistake. Even a minor error could be serious, because in a nuclear plant the tiniest malfunction can develop into a real predicament within seconds. A small change in temperature or coolant level can signal a chain reaction about to go out of control. What happens then?

The first thing many people think of is that the entire plant will blow up, just like a nuclear bomb. However, this cannot happen in any of the commercial nuclear plants operating in the United States today. Their fuel rods contain only enough U-235 to sustain the chain reaction. Each rod would have to hold much more enriched uranium or plutonium before it could react as energetically as even a small nuclear bomb. However, bomblike explosions could be possible in a breeder reactor, like the one that may soon be under construction at Clinch River, Tennessee. Breeder rods would be packed with so much plutonium and other fissionable materials that they could produce a nuclear explosion under certain circumstances. But unless—or until—the United States has a working commercial breeder, nuclear explosions are not a possibility at any power plant in this country.

The LOCA Threat

What is possible in today's nuclear plants is a LOCA, a loss-of-coolant accident. A LOCA could begin in one of several ways. A pipe could break or a valve could stick, allowing coolant to drain away from the core. Or a pump could stop working, uncovering the core.

That would have profound consequences. If the reaction continued without coolant, the core temperature would soar. Even

Safety Systems and an Accidental Test

if the reaction stopped (as it probably would, since loss of coolant means loss of the moderator as well), the fission products in the fuel rods would continue to generate heat. At about 3,700 degrees Fahrenheit (2037. degrees Celsius), the metal casings on the fuel rods would begin to crack and bend. At 5,000 degrees Fahrenheit (2760 degrees Celsius) the uranium pellets would melt. This would mark the start of the worst imaginable nuclear plant accident, a meltdown.

In a meltdown, the fuel rods fall apart. The fuel sinks lower and lower in the reactor, eventually melting the vessel's steel walls. Molten uranium reacts with water, producing steam explosions.

With luck, a meltdown would be only partial. Some, or most, of the fuel would be unaffected. But a meltdown could be complete. It's even conceivable that in a total meltdown, the fuel would go right through the bottom of the reactor building and plunge deep into the earth.

People call this kind of hypothetical accident the "China syndrome," envisioning the molten fuel burning its way straight through the earth to China. Actually the term "China syndrome" is meaningless, since the uranium would go no more than 50 feet (15.24 meters) or so into the ground. At some point, it might hit ground water, producing a tremendous explosion and a massive outpouring of radiation. Radiation leaks could occur during a less disastrous partial meltdown, too, if the steam explosions in the reactor vessel were great enough to damage the containment structure.

The meltdown possibility is scary, but it's not likely to happen, say those who own and operate nuclear plants. The reason for this confidence is defense-in-depth.

Defense-in-Depth and the ECCS

Defense-in-depth begins with such safety features as locating the control room far from the reactor core; automatic scramming; auxiliary manual control; cladding on the fuel rods; the strongest possible construction of the reactor vessel, and so on. But perhaps its most important element is the Emergency Core Cooling System (ECCS).

The ECCS is automatic. If a runaway chain reaction should begin, ECCS pumps are supposed to force water into the core. The water will slow the reaction, giving the core an opportunity to cool. Eventually, workers will bring the plant to a cold shutdown. In a cold shutdown, even the fission products within the fuel rods are giving off very little heat.

The ECCS is vital to plant safety, but it is not the end of the built-in defenses against a huge accidental release of radioactivity. The last line of defense is the containment structure itself.

Even if the uranium fuel melts and steam explosions begin, the containment structure should not be damaged. Its 5-foot (1.5-meter)-thick steel-and-concrete walls are built to withstand larger explosions than those that could be expected during a meltdown. Although the inside of the plant would be filled with radioactive gases and liquids, they should not escape to the outside world. Nuclear Regulatory Commssion rules require that a plant's containment structure be completely leakproof. So even in a meltdown, there would be no breach of containment, no massive release of radiation. Defense-in-depth virtually guarantees nuclear plant safety—or so nuclear industry officials assure us. They add that the same defense-in-depth that protects the public from accidents in the plant also protects it from the results of events outside the plant.

For example, NRC regulations say that plants must be built to weather such natural forces as wind, storms, and tides. A plant must also be able to stand up against earthquakes of an intensity greater than that which would normally be expected in the area. The industry points with pride to its earthquake record. On one occasion, four Japanese plants operated without a pause during a quake that measured 7.5 on the Richter scale. (On this scale, the very strongest earthquake rates a 10.)

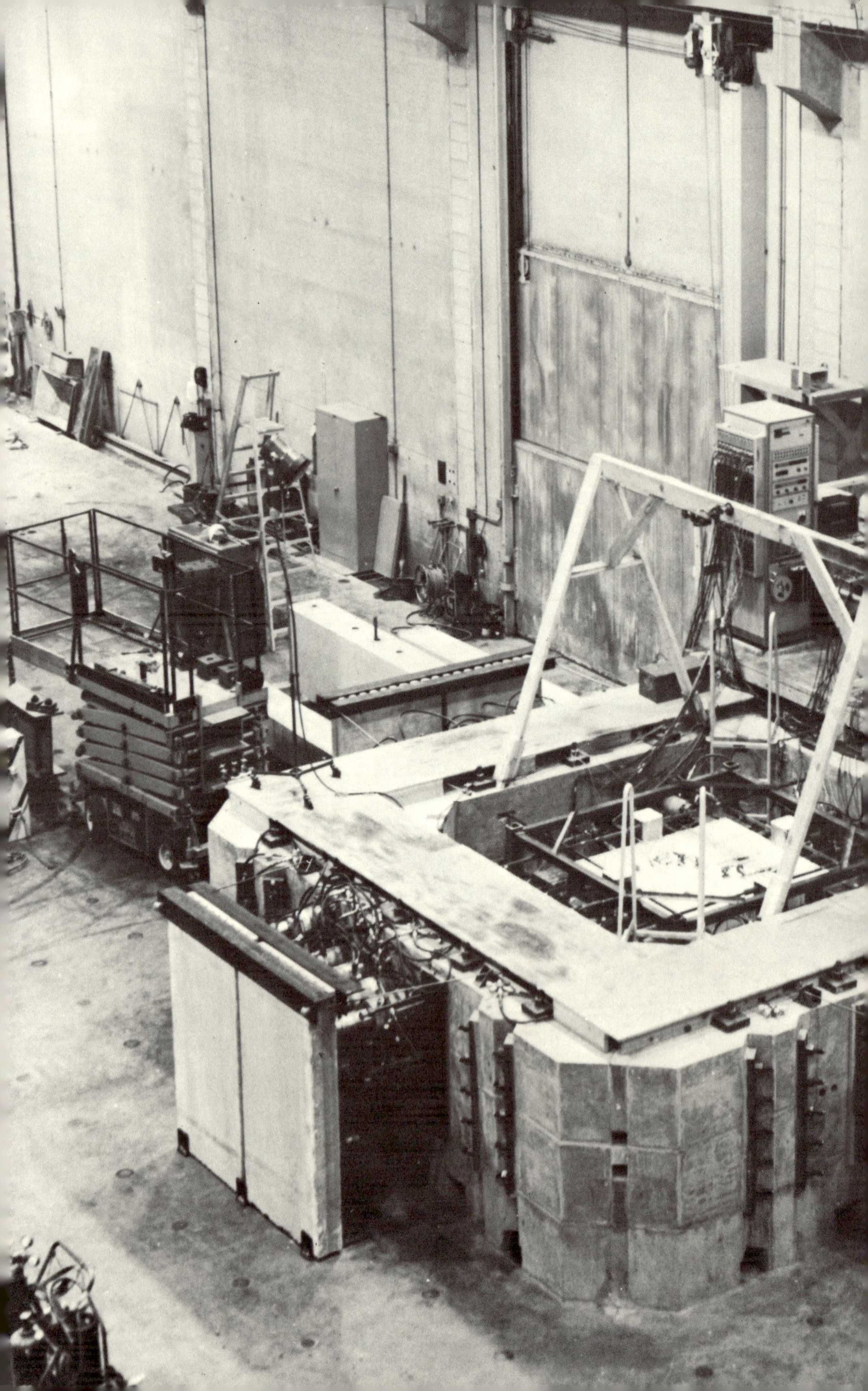

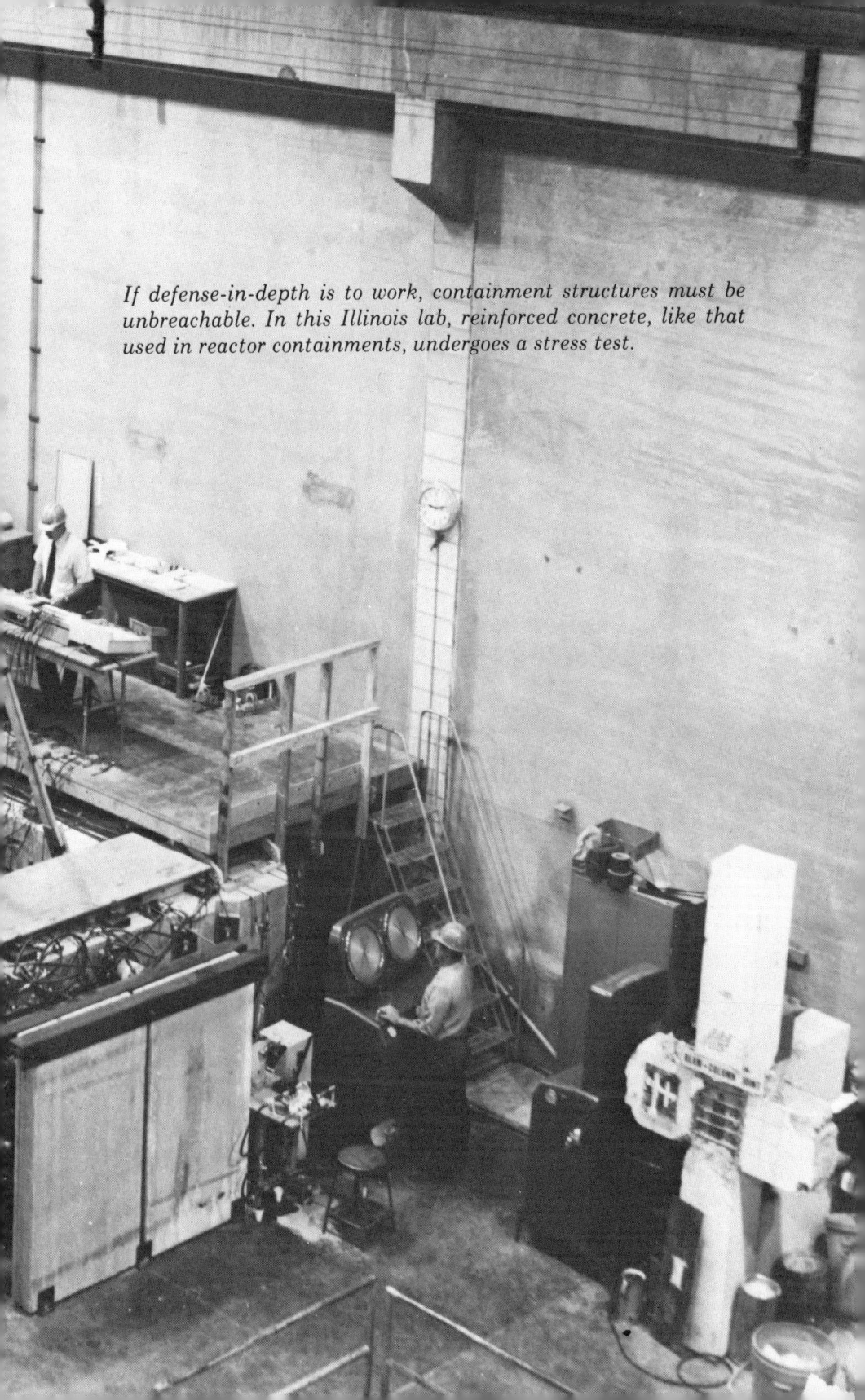

If defense-in-depth is to work, containment structures must be unbreachable. In this Illinois lab, reinforced concrete, like that used in reactor containments, undergoes a stress test.

Workers inside a nuclear plant on the Maine coast didn't even feel a force-4.0 earthquake centered just 10 miles inland.

People in the industry are delighted with the results of such natural tests of plant safety, but government regulators also want to see the results of scientific tests of safety features. Accordingly, the ECCS has been extensively tested to see how well it works. Besides that, there have been numerous studies of what could go wrong in a plant, and pages and pages of computer analyses on how best to handle various emergency possibilities.

Test Completed—Results Uncertain

What have the analyses and tests shown about reactor safety and the ECCS? How inadequate both are, answer critics of nuclear power. The first ECCS tests were conducted on small-scale models at government laboratories in 1966 and 1967. Researchers at the labs agreed that they did not have enough information to say definitely whether or not the ECCS really worked. The final report of the research team at the Oak Ridge National Laboratory in Tennessee called the adequacy of the ECCS "speculative."

Later tests at Oak Ridge were equally disappointing. They showed that during a nuclear accident, the fuel rods would swell. Their enlargement would obstruct the flow of emergency coolant. Experiments at the National Reactor Testing Station in Idaho showed much the same result. What's more, computer analyses indicated that the core temperature during an accident would be higher than first believed, making a meltdown more likely.

In 1970 and 1971, scientists performed another series of tests on scale-model ECCSs. Again, the systems failed. Pipes broke, draining the coolant from the core. But the testers still felt that they lacked enough basic information about the ECCS to judge its efficiency. In August 1971, researchers at the government's Idaho lab drew up a list of twenty-eight areas in which information was "inadequate," "missing," "uncertain," or "unverified."

The nuclear industry has a different perspective on the test results. Its officials maintain that the unsuccessful scale-model tests should not be considered because none was of a real ECCS. They emphasize that after the 1966–67 tests, federal regulators gave the green light to the ECCS. Furthermore, they say, expert testimony at government hearings between January 1972 and December 1973 vindicated the ECCS and plant safety overall. The American Nuclear Society cites a poll showing that after the two years of hearings, forty-four out of forty-six "technical participants" questioned thought the ECCS was safe enough. A government report released in 1978 says that 97 percent of the staff of the NRC Office of Nuclear Regulation believe that nuclear plants are as safe as, or safer than, they need to be. A new series of tests, begun in December 1978, also demonstrates ECCS efficiency, the NRC says.

Who's right? It's hard to tell, because each side can quote facts and figures to back up its own point of view. To try to sort out the confusion, government regulators ordered, in 1972, a complete study of reactor safety. Named to head the study group was Norman C. Rasmussen, a professor of engineering at the Massachusetts Institute of Technology. The group's report took three years to compile, and cost $4 million. Although its formal name is the Reactor Safety Study, it is usually referred to as the Rasmussen Report.

After examining the evidence, the study group turned in a document that played down the possibility of a serious reactor accident. A total meltdown, according to Professor Rasmussen's calculations, has only one chance in 20,000 of occurring at any given reactor in any given year. Put another way, if the country had one hundred operating reactors, we would see one meltdown every two hundred years.

The Rasmussen Report also minimized the human injury that

would result from a nuclear accident. In his first draft, Rasmussen predicted that about 2,300 people would die immediately in a nuclear plant disaster. About 5,600 would become ill. The economic cost of a severe accident would amount to $6.2 billion.

The report came in for quick rebuttal. The Union of Concerned Scientists claimed that it underestimated the chances of a serious accident by a factor of 30. A committee of physicists representing the American Physical Society criticized the report because it did not consider the health effects of low-level radiation released in an accident. Responding to the criticism, Rasmussen raised his figures to 3,300 immediate deaths, 45,000 immediate injuries, 45,000 long-term cancers, and 5,100 long-term birth defects. He increased the dollar cost of an accident to $14 billion.

The changes did not satisfy the critics. The UCS points out what it considers serious oversights: The report failed to discuss fire damage at a nuclear plant. It did not examine what might happen if a plant had to be abandoned during an accident. It did not adequately look into the problem of incorrectly installed or poorly maintained equipment. Addressing itself only to reactor safety, it said nothing about radiation hazards at other points in the fuel cycle.

Even the NRC had its doubts. In January 1979, the commission rejected the document. It questioned Rasmussen's assertion that he had, if anything, overstated the likelihood of a reactor accident. Nevertheless, the NRC said that serious accidents were a remote possibility and people should not worry about them. For three months, Americans took that advice. Then came Three Mile Island.

Accident

The accident in the Number 2 reactor began at about four o'clock on the morning of Wednesday, March 28. A valve failed, slowing down the flow of coolant into the reactor. The core temperature began to climb. In the control room, an alarm klaxon blared. Red lights flashed.

Instantly, computerized devices took over and dropped the control rods into the core. The reactor scrammed. However, fission products in the fuel rods were still producing huge amounts of heat. The ECCS switched on automatically to provide coolant.

But the ECCS was switched off too early because of a confusing instrument reading. That major safety violation, which reduced the amount of secondary cooling water in the core, had gone unnoticed—until the pumps were needed.

Lacking sufficient cooling water, the core continued to heat up. Now, according to the defense-in-depth principle, there should have been another automatic safety device on hand to cope with the situation, and indeed there was. A valve at the top of the reactor opened, allowing some of the core's heat to escape in the form of steam.

Unfortunately, another problem arose at once. The valve at the top of the reactor did not close, as it ought to have, after allowing the steam to vent. It stayed open, and thousands of gallons of steamy coolant billowed out. The water level at the core dropped further. By now, about five minutes into the accident, the tops of the fuel rods were exposed, and the temperature was still mounting. A LOCA was underway. TMI-2 was on the verge of a meltdown.

And the four on-duty operators in the control room were utterly unaware of it. As far as they knew, a malfunction had occurred and had quickly been corrected. A gauge in the control room clearly showed that the top-of-the-reactor valve had closed and that the coolant had been replaced. Their eyes on the gauge, the men had relaxed and shut off the ECCS by hand.

A minute later, the core temperature started upward once more. At nearly 3,000 degrees Fahrenheit (1648 degrees Celsius) the fuel rods were beginning to buckle. The ECCS

Safety Systems and an Accidental Test **77**

When something goes wrong at the core, the reactor should scram and the control rod automatically drop into place. That's what happened at Three Mile Island, but human error permitted a serious accident to occur there. This photo, taken at a different plant during refueling, shows (foreground) the shafts that hook onto the control rods, allowing them to be raised and lowered.

switched itself on again, and again an operator turned it off. It was not until 6 A.M. that a plant supervisor discovered the open valve atop the reactor.

As the stuck valve was closed, highly radioactive water began spilling onto the reactor-room floor. A pump sucked up the water and transferred it to containers in another part of the plant. There it spilled again. Radioactive gases welled up.

Plant operators had to get rid of the gases. Over the next hours, they vented small bursts of radioactivity into the air above the plant. Later, plant personnel released thousands of gallons of mildly radioactive water into the Susquehanna River.

Inside the control room, the situation was chaotic. A hundred separate alarm bells were ringing in a deafening chorus. Operators ran from one set of controls to another. Early in the afternoon, a small amount of hydrogen gas exploded in the reactor room. The explosion did not damage the building's outer shell, however, and containment was not breached.

By Friday, March 30, the plant operators had still not brought the plant to a cold shutdown. The Nuclear Regulatory Commission decided to send in a team of nuclear experts to see what they could do. The experts could not examine the reactor at first hand; to approach it would have meant death within thirty seconds. But by studying the control instruments and analyzing what had gone on over the previous forty-eight hours, the experts managed to get a picture of the situation. They believed that a mass of hydrogen gas was trapped at the top of the reactor vessel. This "bubble," as some called it, seemed most likely to act in one of two ways. It might explode, possibly wrecking the containment structure and spewing radioactivity far and wide. Or it might displace what coolant remained in the reactor, leading to a total meltdown.

That Friday evening was a grim one in Pennsylvania and across the country. Hundreds of local families left their homes to seek shelter with friends or relatives in other towns. Those with no one to turn to filled schools, churches, and other public buildings outside the immediate area of the damaged plant. Elsewhere in the state, and around the world, people waited to learn whether nuclear technology could save the day.

Nuclear technology did not. A simple natural phenomenon—the fact that hydrogen gas dissolves in water—did. The liquid coolant began to absorb the hydrogen. By Monday, the gaseous mass was shrinking, and within a week it was gone. Those who had left their homes returned; the accident was over.

Investigation

Investigation of its cause began at once. The President named a special commission to look into the events at TMI. He appointed John G. Kemeny, president of Dartmouth College in New Hampshire, to head the group. Other commissioners included a businessman, a labor leader, a conservationist, engineers, university people, and a housewife from Middletown.

Throughout the summer and into the fall, commission members studied the evidence. Much of their studying had to be done at second hand, since the reactor room was still off-limits. But the group did examine the business records of the electric utility company that operated the TMI plant. The utility, Metropolitan Edison, was part owner of the plant.

The commissioners also researched the type of equipment used at TMI and compared its operation there with the performance of similar equipment at other plants. Finally, investigation of the events of late March and early April told the commissioners what they needed to know. They submitted their report to the President in October.

The report contained both praise and blame for nuclear power plants and for the way they are run. On the plus side, the commissioners concluded that equipment, including the ECCS and other safety features, were satisfactory. "Overall, their equipment was very good," Kemeny told reporters. Commissioner Thomas Pigford, of the department of nuclear engineering at the University of California, concurred. "Equipment failures were not the . . . cause of the TMI-2 accident," he wrote. "The accident was, in fact, a demonstration that the equipment is effective." Pigford's and Kemeny's comments were made on the basis of the fact that the reactor automatically scrammed at the first sign of trouble, and that the ECCS switched on several times as the reactor overheated

Blame for the accident rests, according to the report, on human failures. The control-room personnel were poorly trained, and they did not understand what was happening at the core. In their ignorance, they repeatedly turned the ECCS off. It took two hours for someone to realize that a stuck valve was letting coolant rush from the core. Lack of training, said Kemeny, left the operators "unprepared to deal with something as confusing as the circumstances in which they found themselves."

Of course, the people in the control room can hardly be criticized for their poor training. The blame for that was placed on Metropolitan Edison and the NRC. The former was criticized for failing to provide thorough training, and the latter for failing to require it.

More blame fell on Met Ed. The plant was shoddily maintained, the commissioners found. When they toured TMI-2's sister reactor, TMI-1, they saw that leaks from broken parts had allowed mineral deposits to build up on valves and pipes. Reported the commission, " . . . boron stalactites more than a foot long hung from the valves, and stalagmites had built up from the floor." Little wonder that a valve might stick!

The possibility that a valve might stick should have been foreseen anyway. More than a year before the accident, a nuclear plant in Ohio suffered two separate breakdowns due to a jammed valve. The plant's reactor had been designed and built by the same company that produced the TMI reactor, and the valve that stuck in Ohio was exactly like the one that stuck at TMI. Operators at the Ohio plant managed to control the situation both times, and they informed the reactor manufacturer of the problem. It took the manufacturer one full year to pass this information along to other plants with similar equipment. The manufacturer's warning reached TMI in April 1979—seven days after the accident began there. After that, TMI's owner announced that it was suing the manufacturer for $500 million. In a separate action, the utility filed a claim for $4 billion against the NRC on the ground that the commission failed to convey a warning about the valve's tendency to stick.

The Kemeny Commission presented a lengthy report about what had gone wrong at TMI. But put simply, their conclusion was: Nuclear technology and equipment are reasonably reliable—certainly more reliable than are the people who own, operate, and regulate nuclear plants.

It's hard to disagree with the commissioners' verdict, at least as far as the mismanagement at TMI is concerned. What about the nation's other nuclear power plants and related facilities, though? Are they also subject to accidents due to human carelessness and mismanagement?

7 OTHER ACCIDENTS, OTHER THREATS

The accident at Three Mile Island may have been the worst ever at an American nuclear installation, but it was far from being the only one. Almost equally dramatic, and for a time equally threatening, was the 1975 fire at the Brown's Ferry nuclear plant near Decatur, Alabama.

The Brown's Ferry plant is owned and run by the Tennessee Valley Authority (TVA). On March 22, 1975, two workers were placing a rubbery material around electrical cables just below the control room. The purpose of the packing was to block air leaks. To see whether the leaks were sealed, the workers held a lighted candle close to the cables. If the flame remained steady, there was no leak.

Suddenly, one of the men held the

candle too near the rubber. The packing burst into flame, and the fire quickly spread along the cables and into the reactor room. It burned for seven hours before plant workers, who lacked adequate fire-fighting equipment, got it under control.

The fire did a great deal of damage. It destroyed the cables that carried electricity to run the reactor's cooling system. The ECCS could not take over—it was disabled, too. Its electrical cables ran alongside those of the primary system. The fire knocked both systems out of commission simultaneously. Without coolant, the core began to overheat.

Plant operators kept their heads. They located some undamaged cables and fashioned a makeshift connection to an auxiliary pump. With this arrangement, they kept the core temperature at a safe level until they could bring it to a cold shutdown.

Unlike the TMI accident, the Brown's Ferry fire was not investigated by a presidential commission. Had it been, a similar pattern would have emerged. Nothing was wrong with the equipment. Once the auxiliary pump was working, the accident was over. But human error triggered the accident to begin with.

Human beings laid the two sets of cables so close together that damage to one meant damage to the other. Human beings used flammable rubber packing and tested that packing with fire. Human beings failed to provide fire-fighting equipment within the plant.

Human error of a different kind was responsible for an October 1978 mishap at a federal government laboratory in Idaho. The problem began when uranium that was being processed started to go critical. The supervisor on duty should have realized what was happening and brought the chain reaction under control at once.

But this was in October—World Series baseball time—and the supervisor was watching the sixth game between the New York Yankees and the Los Angeles Dodgers. The uranium started to go critical in the seventh inning, with Yankee Roy

White on first and Reggie Jackson at bat. Engrossed, the supervisor had no inkling that an emergency was developing. Other operators on duty also failed to notice what was happening.

Cra-a-a-ck! Clang! Clang! A homer for Jackson. And for the men in the Idaho control room, the warning cry of radiation alarms. The men dashed from the room, activating machinery to halt the reaction as they fled.

Much more serious was the accident twelve years earlier at the Enrico Fermi reactor near Detroit, Michigan. What made this accident especially threatening was that the Fermi plant housed an experimental breeder. As the men on duty knew, a breeder, unlike a conventional reactor, can experience a bomblike nuclear explosion. It's not difficult to imagine the men's horror as they realized, shortly into the accident, that the core was starting to melt. The operators acted quickly and competently, however. They stopped the meltdown and eliminated the chance of a blowup. Weeks later, when workers began dismantling the core, they found the cause of the accident—a scrap of metal jammed against a coolant duct. The scrap had kept coolant from reaching the core as it heated.

On examination, it was discovered that the scrap was a piece broken from a safety device installed in the reactor seven years earlier—and forgotten. Even a search of old plant records and blueprints revealed no mention of its presence.

Yet another potentially serious accident seems to have been narrowly avoided at the Indian Point plant in New York State late in 1980, when 100,000 gallons (378,500 liters) of Hudson River water overflowed a sump pump and flooded into a pit beneath the reactor vessel. At first, the plant's owners—officials of Consolidated Edison Company—considered the accident "a pumping problem." They failed for three days to inform the NRC of what was going on. When at last they did admit that the accident had occurred, they also announced that repairs would take at least seven months.

A Future of Accidents?

Human mistakes have led to nuclear accidents in the past. Will they cause more in the future? The answer may well be yes. Serious safety violations keep cropping up at nuclear plants around the country.

At a Michigan plant, workers neglected to shut the valve on a pipe leading from a reactor containment building to the outside. The pipe, 4 inches (10.16 cm) in diameter, was left open between April 1978 and September 1979. If a TMI-type accident had occurred at the Michigan plant any time during those seventeen months, radioactivity would have poured out of the building.

Of twenty-eight plants reviewed by the UCS in 1978, twelve had wiring systems like those at the Brown's Ferry installation. Cables for the primary cooling system and for the ECCS were laid side by side, inviting a situation in which damage to one would mean damage to the other. Apparently, the owners of the twelve plants had learned little from what happened at Brown's Ferry. The UCS notes that the owners did agree to correct the problems—but not until the end of 1980.

Nuclear Proliferation, Nuclear Terror

Human beings may well cause future nuclear accidents by mistake. They could also bring about a nuclear disaster on purpose.

In 1974, India took a step in that direction by exploding its first nuclear bomb. Materials for the bomb came from a nuclear reactor that Canada had sold to India. Canada made the sale on the condition that the reactor be used only for nonmilitary purposes. But how were the Canadians—or anyone else—supposed to enforce that proviso? Now, the United States has agreed to sell 38 tons (34.5 metric tons) of enriched uranium fuel to India. Many experts believe that much of this fuel may be used to produce nuclear weapons.

Radioactive materials are constantly being shipped throughout the United States and to other countries. This NRC inspector checks to see that containers of fresh reactor fuel, headed for Japan, are contamination-free.

The governments of the United States and Canada have been open about selling nuclear materials to India. Not all governments are so straightforward. And, of course, the governments that buy reactors in order to fabricate bombs on the sly will do so secretly. No one will know who's selling to whom. No one will know how many countries, large and small, have nuclear weapons. It's a sure bet, though, that before too long, many nations—some with dangerously unstable leadership—will be capable of nuclear war. This is the possibility that disturbed President Carter so much that he imposed a moratorium on spent-fuel reprocessing.

Nuclear terrorism is another possibility, or so some people believe. Political terrorists could break into a nuclear plant and threaten to bring about a meltdown and a massive release of radioactivity unless certain demands were met. Terrorists might raid a reprocessing plant and steal plutonium. This could be used to make a nuclear bomb by anyone with a flair for college physics. Or terrorists might sneak the material out, bit by bit, until they had enough for a weapon.

Other Accidents, Other Threats

According to nuclear power proponents, however, nuclear terrorism and sabotage are extraordinarily remote possibilities. To start with, the industry points out that the United States military has been in the nuclear business since World War II. In 1979, the government had a stockpile of approximately 30,000 nuclear weapons. But no terrorist has ever held up a weapons fabrication plant and threatened nuclear destruction. None has ever run off with uranium or plutonium and constructed and detonated a bomb.

Why not? One answer has to do with security. Military nuclear installations are so well guarded that terrorists and saboteurs cannot get in.

Security at civilian nuclear facilities is just as tight, industry officials claim. Armed men stand watch at every nuclear power plant. They would be on hand to protect reprocessing plants and breeders, too. Anyone seeking to enter a plant is checked out by security forces ahead of time. Visitors must prove that they have legitimate business inside. At the plant entrance, they pass through one apparatus that "sniffs out" explosives and another that detects any hint of concealed weapons. As they leave, they must submit to a radiation check that reliably indicates whether they are carrying radioactive materials.

Another reason why terrorism and sabotage are not threats, nuclear advocates contend, is that people in their right minds would not mess around with radioactivity. The terrorist who contrives a nuclear accident will be killed or injured just as surely as everyone else in the vicinity. The person who steals uranium or plutonium and hides it around the house is going to be the first to feel its radiation effects. Building a nuclear bomb is a highly technical undertaking that no one can tackle without specialized knowledge. It's unlikely that anyone would even try.

The arguments sound convincing, but there are holes in them. A look at the record shows why.

Security at nuclear plants isn't always what it's cracked up to be. In 1979, the Associated Press news service conducted a four-month investigation of security at the Indian Point plant in New York. Among charges made by guards at the plant were:

· They were poorly trained. Company records were falsified to indicate that guards had passed training courses they never even took.
· Some guards couldn't use their guns. Weapons were issued to guards who had failed one marksmanship test after another.
· Warning alarms on some fences were disconnected.
· Keys to off-limits areas were passed casually from person to person.

One Indian Point guard summarized the situation as he saw it. "It would take one commando or one well-trained wacko to go in, inflict heavy damage, and leave without being detected," he told an AP reporter.

Indian Point is not unique. Not long ago, the Oregon state police broke up a drug ring operating among security guards at a nuclear plant in that state. How reliable would those guards have been in a crisis? Were they on drugs themselves? Would they have been susceptible to blackmail?

Nuclear Theft

Even the NRC is concerned about security. In 1976, the commission's director of safeguards wrote that he feared plants were not sufficiently protected against theft. Later that year, another government agency reported that "tens of tons" of materials that could be used for weapons could not be accounted for. The materials had disappeared from thirty-four plants that were carrying out projects for the federal government. More recently, officials at a uranium-fabricating plant in Tennessee realized that they were missing 22 pounds (9.98 kg) of bomb-grade uranium. An intensive three-month search failed to turn up the missing material.

Nuclear officials believe that this, and other missing uranium, is unlikely to reappear in a bomb. They may be wrong. Nuclear bombs aren't as difficult to build as nuclear advocates would have us believe. To prove the point, a public TV station asked a graduate student to try to construct a bomb. The student did. One expert who looked at the weapon said it had a good chance of working. Recent newsmagazine articles have outlined exactly how someone can go about making a nuclear bomb. Warns one leading American physicist, "If a terrorist gets his hands on plutonium oxide, I think the probability exceeds 25 percent that he can convert it to an atomic bomb in short order."

The industry's assumption that people in their right minds would never attack a nuclear installation isn't reassuring, either. Who's to say that terrorists *are* in their right minds? Even those who are clinically sane may be carried away by their convictions. Scores of modern terrorists have demonstrated their willingness to die for a cause. During World War II, thousands of Japanese kamikaze pilots sacrificed their lives on suicide missions against American ships in the Pacific Ocean.

It's even possible that saboteurs might strike at a nuclear plant in order to *prevent* accidents. That's what two control-room trainees at a station in Surry, Virginia, did. The two had complained repeatedly about what they saw as safety defects at the plant. To dramatize how easily anyone could get in and damage the reactor, the men crossed over to the hot side of the plant and dumped caustic soda over the fuel rods. The men were later convicted of intentionally damaging a public utility, but they hardly minded. They had made their point.

Nuclear workers disable a reactor. Bomb-grade materials vanish. Plant security is questionable. Operator training is poor. Safety precautions are overlooked. Backup systems are carelessly installed. The nuclear safety record does not seem very impressive.

That's not so, nuclear proponents protest. On the contrary, they say, the industry's record is excellent. Safety oversights and a few near-accidents are only part of the story. It's not fair to concentrate on a small number of problems and to ignore the thousands and thousands of days that nuclear reactors have produced electric power without the slightest sign of trouble. The fact that no one has ever been killed by radiation poisoning at a commercial nuclear plant cannot just be brushed aside.

The harshest critics of nuclear power have to agree that so far no massive amount of radiation has been known to escape from a nuclear plant. Even at TMI, containment worked. Federal radiation experts say that only minute amounts of radioactivity were released during the accident. (However, more was released during the long process of cleaning up the damaged reactor, and two years after the accident there were reports that wild animals in the TMI area were leaving radioactive droppings.) Only one or two long-term cancer deaths are now expected to result from the releases.

Looked at this way, the TMI accident is almost reassuring. The safety systems did work. Containment worked too, at least during the days of the accident itself.

But TMI did more than demonstrate that containment and the ECCS work. The scare reminded Americans that a major nuclear accident would be no ordinary accident; it would unleash deadly poisons, poisons that would kill and maim, that would infect the food chain for centuries, that would contaminate land, air, and water—and that would do it all without giving any warning of their presence.

The accident also brought home the fact that nuclear power does not involve technology alone. Nuclear power involves the people who build and buy the technology and those who guard it and operate it. And no matter how brilliant and ingenious nuclear technology is, or becomes in the future, it will never be any better than the human beings in whose hands it lies. Or as John Kemeny summed it up to reporters, "The plants are safe; it's the people who aren't safe."

8
REGULATION OR PROMOTION?

"It's the people who aren't safe."

Is there any way to change that? The men and women who served on the Kemeny Commission think so. Their solution: better government regulation of the nuclear industry. The commission's final report recommended abolishing the five-member Nuclear Regulatory Commission and replacing it with a single "director of nuclear regulation."

Having one regulator instead of several, the Kemeny group believed, would mean more efficient policing. "We felt that when you have a collegial body, you delay decision-making while searching for the lowest common denominator of agreement," one commissioner explained. A single director, he went on, "can't pass the buck in an emergency."

The Kemeny Commission is not alone

in its criticism of the NRC. Antinuclear activists think the NRC is too lax in its dealings with industry problems. Some commissioners and staffers have worked in the nuclear industry or for a public utility, or expect to work in the conventional power or nuclear industries after leaving the NRC. Close ties with nuclear power interests can lead people at the NRC to put industry concerns ahead of the public interest and public safety.

Within the nuclear industry, the feeling is different. Here, the complaint is that the NRC is too fussy about safety requirements and too slow to license new plants. Industry officials point out that it now takes from twelve to fourteen years to get a plant designed, built, and put into operation. Plans must be examined, and safety features inspected. The company building the plant is required to show that the proposed location is a safe one. The company must take into consideration what effect the plant will have on the local environment—land, air, water, and plant and animal life.

The NRC isn't the only regulator body that must be satisfied. Altogether, as the nuclear industry points out, there are over forty federal agencies that have a say in energy production and electric utility operations. State and local agencies may have their own regulations, which could necessitate postponing plant construction month after month.

Every postponement is expensive. According to the industry, the cost of getting a nuclear plant on line increases by $10 million for each month of delay. Consequently, the nuclear industry is demanding changes in licensing and regulatory procedures.

Antinukers want changes, too. So does the Kemeny Commission. The government's role in nuclear power seems to be under attack from nearly everyone. Why?

The Atomic Energy Commission

It all goes back to just after World War II, when Congress passed the Atomic Energy Act of 1946. This act set up the

License application forms, safety analyses, and other documents pour into NRC headquarters by the truckload whenever new plant construction is proposed. As many as forty copies of a single report may have to be sent to the various government agencies involved in safety reviews. The nuclear industry would like to cut down on paperwork and speed up the license application process.

Atomic Energy Commission (AEC), and charged it with studying ways that nuclear power might be used for peaceful, commercial purposes.

The five-member AEC, aided by its staff of technical experts, undertook the task with enthusiasm. The nuclear dream was bright in 1946, undimmed even by the awareness that a single atomic bomb could wipe out most of a city. After all, the war was over, and nuclear energy could be put to use to benefit mankind.

After eight years of consideration, the AEC informed the President and Congress that enough was known about nuclear energy to make it commercially feasible. The AEC urged that the country's electric utility companies start building and operating nuclear plants to generate power.

In response, Congress passed another atomic energy law, the Atomic Energy Act of 1954. The new law gave the AEC the authority to oversee the infant industry—to grant each nuclear facility its license to operate, and to write the rules under which the plants would be run.

Now the AEC had a double job. Under the 1954 act, it was supposed to *regulate* the nuclear industry. But under the 1946 act, it was supposed to *promote* the industry and give it as much support as possible.

That was an impossible assignment. It's like putting the commissioner of baseball in charge of getting Americans to spend less time at the ballpark, or making a pacifist a five-star general. No one can promote something and control it at the same time.

Perhaps the commissioners didn't realize that. They went right ahead with their paradoxical mission, regulating and promoting simultaneously. Inevitably, they did more of one than of the other. As it happened, what they did more of was promotion.

This is not surprising. Promotion had been the AEC's first responsibility; the regulation part had come later. More important, most of those connected with the AEC had been thoroughly committed to nuclear energy long before they took their new jobs. The AEC was headed and staffed by some of the country's leading nuclear scientists and engineers. They were knowledgeable about nuclear energy, and they felt comfortable with it. Many of them had built and run their own atomic piles, and they knew exactly how to control the reaction. They had worked with ionizing radiation, learning how to deal with it safely. They had handled radioactive wastes in their labs.

One early AEC chairman, for instance, was nuclear physicist Glenn T. Seaborg. Seaborg, who had worked with Enrico Fermi at the University of Chicago, was the codiscoverer of plutonium. He wasn't intimidated by nuclear technology, and he knew how to protect himself and those around him from radiation. Naturally, the making of safety rules was not uppermost in Seaborg's mind. He was primarily interested in encouraging nuclear development and in attaining his goal of an energy-rich America "set free" by the plutonium breeder. When, in 1962, Seaborg submitted a lengthy report on nuclear power, he made no reference at all to radiation dangers. He did include a list of "Objectives for the Future," but safety was not among them.

Other commissioners who shared Seaborg's point of view believed that nuclear accidents were so unlikely that they need hardly be considered in making rules for the industry. One result of this attitude was that federal regulators permitted the building of some nuclear plants in what most people now agree are inherently unsafe locations.

Lax Siting Rules

New York's Consolidated Edison Company, for example, got the go-ahead to construct a three-reactor plant at Indian Point, just up the Hudson River from New York City. The plant is 20 miles (32.18 km) from the Bronx, the city's northernmost borough. It is 40 (64.36 km) miles from Times Square. More than 20 million people live within a 50 (80.45-km)-mile radius of

Indian Point. The AEC's faith in reactor safety is also responsible for the siting of Illinois' Zion plant reactors, 41 (66 km) miles from Chicago and 40 from Milwaukee.

An accident at Zion or Indian Point would cause massive panic and disruption. Even if no one were harmed by radiation, hundreds—even thousands—might be killed or injured in traffic jams created by fleeing citizens. Heart attacks and strokes would claim many more. Where would the millions of nuclear refugees go? Public buildings in safe places wouldn't be big enough to hold them all. How would they be fed? Who would supply medicines when diseases began—as they would begin—sweeping through the tightly packed living quarters? How many people, especially children and the elderly, would die of those diseases?

The AEC never had to try to answer such questions, because it neglected to ask them. That neglect has been a boon to the nuclear industry. A company like New York's Con Ed naturally prefers to build its power plants as close to New York City as possible. Being close cuts down on the cost of transporting electricity once it's generated. Less costly equipment needs to be installed. Fewer expensive breakdowns will occur along the line.

The AEC's lax siting rules have saved money for the industry even where population density isn't a problem. In 1973, the Virginia Electric & Power Company (VEPCO) was constructing its North Anna nuclear power plant in the town of Mineral, almost 50 miles (80.45 km) from Richmond. Suddenly a group of Virginia environmentalists discovered that the plant was being built directly over a geological fault.

Charging that it is unsafe to place a nuclear reactor on the exact spot where earthquake potential is greatest, the environmentalists tried to get the AEC to revoke its permit for the North Anna plant. The AEC refused, although it did eventually fine VEPCO for attempting to cover up evidence that the fault existed. But the AEC spared the utility the far greater expense of giving up the North Anna site and finding a safer location for its reactor. The North Anna plant is in operation today.

Lax Safety Rules

Once a plant is built, it must be equipped. In the matter of equipment, too, regulations have been eased to help the utilities meet them.

We've already seen proof of this. The AEC approved use of the ECCS in the late 1960s even though test results showed that the system was inadequate. AEC approval of the faulty system was no oversight; the commissioners and their staffs *knew* that the ECCS was not satisfactory.

In 1971, a technical adviser to the AEC wrote that reactor safety systems did not give "sufficient assurance to provide a clear basis for licensing." But licenses went right on being issued.

The next year, the AEC's Advisory Committee on Reactor Safeguards urged that the commissioners demand "significant improvements" in the ECCS. The commissioners did no such thing.

The year after that, the Federation of American Scientists, having reviewed existing ECCS studies, asked the AEC to cut back on nuclear plant operations. The AEC did not.

The ECCS was just one unproven safety system. There have been others. Often, the AEC was aware of certain problems in older plants, and knew that those deficiencies had been corrected in newer ones. But the commissioners did not require older plants to upgrade their equipment. This, too, saved money for the utilities. So did the fact that the AEC did not establish fire safety standards for nuclear plants. Apparently it never occurred to the commissioners to do so—until after the events at Brown's Ferry. Nor did the AEC attempt to regulate the practice of checking flammable materials for leaks in the easiest and cheapest way: with a candle.

Assuming that nuclear power was basically safe, the commissioners also assumed that as problems cropped up in the industry, scientists and technologists would promptly solve them. That's the assumption they made, for instance, about reprocessing and the storage of radioactive wastes.

In the late 1950s and early '60s, when commercial nuclear power was just getting off the ground, reprocessing was not an immediate necessity. The country had only a few small nuclear plants, and plenty of uranium to fuel them. Waste storage was not a problem yet, either. Few wastes had been produced.

Now the picture is different. The utilities that own nuclear plants are searching for a place to dump their radwastes. They are also looking ahead to a day when, without reprocessing, they will run short of uranium. But so far, reprocessing has proved neither safe nor profitable in the United States. No one has figured out what the best means is of safely shipping and storing radioactive garbage. The AEC's serene confidence that problems would automatically be solved as they arose has allowed the nuclear industry to use up scarce resources and produce deadly wastes for twenty-five years—and left us facing unsolved problems.

Having set up its lax rules, federal regulators proceeded to enforce them gently—if at all. The AEC largely accepted the industry's recommendations for the quantity of radiation that plants should be allowed to release, for instance. And although plants are supposed to report each release, there is no established way of making sure that they do.

Plants have not been rigorously inspected, either, nuclear critics claim. In 1979, there were only twenty-two regular federal plant inspectors spread around seventy-two operating plants. When a plant is checked, it's rarely a surprise to plant owners and personnel. Says a former guard at the Indian Point plant, "We know ahead of time every time anyone comes through. They call all the guards and tell us to look sharp by broadcasting the code message 'Signal 30.'" An inspection that employees can prepare for amounts to no inspection at all.

Even when inspectors did spot a problem—or had it pointed out to them, as in the case of VEPCO's North Anna plant—the consequences to the plant's owners could scarcely be called severe. VEPCO's fine for lying about the geological fault was $32,500.

Evacuation Vacuum

Another gap in federal regulations concerns the lack of evacuation plans for local inhabitants in the case of a nuclear accident. The AEC never required anyone, not a plant's owners, not state or local police, not civilian emergency groups, to think about the possibility of an emergency.

What can happen without a plan was demonstrated at TMI. Although no general evacuation was announced, the governor did suggest that pregnant women and preschool children leave the area. Even that was enough to disrupt Middletown. Lines formed at gas stations. Roads were crowded. Telephone callers got nothing but busy signals. In nearby Harrisburg, a siren went off, convincing some that a wholesale evacuation had been ordered. Men and women rushed from restaurant tables into the streets. A few hospital employees ran out on their jobs. Terrified prisoners screamed that they were being left to die behind bars.

At TMI, wholesale evacuation wasn't necessary. But somewhere, sometime, it may be. For state and federal officials to allow nuclear facilities to go on line without workable evacuation plans seems little short of negligent. In fact, government regulation of nuclear power overall has been inadequate. Time and again, regulation has bowed to promotion.

By 1974, this fact was becoming apparent to more and more people. That year, Congress passed the Energy Reorganization

Act, which abolished the AEC and split up its two-part mandate. The job of developing and promoting nuclear energy went to a new organization, the Energy Research and Development Administration (ERDA). Not long afterward, ERDA and parts of several other agencies were reorganized into the federal Department of Energy (DOE). The responsibility for regulating the nuclear industry was given to the newly created Nuclear Regulatory Commission.

The NRC and Business as Usual

If Congress's purpose was to set up a strong and safety-minded regulatory body, it failed. Most of the men and women who got jobs on the new NRC had previously worked for the AEC. Their titles had changed, but their ideas and attitudes had not. One of their first acts as NRC members was to adopt the old AEC rules, regulations, and procedures. They made almost no changes in AEC practices.

Replacing the AEC with the NRC has not accomplished what some people hoped it would. Many of the regulatory lapses we've seen in this chapter occurred after the NRC took over. At the same time, the nuclear industry itself is displeased by the NRC's performance. The commission's clumsy, drawn-out licensing procedures are doing their part to raise the cost of building new plants. That annoys investors. No one wants to wait twelve or fourteen years to start getting a return on an investment. Already, investors are beginning to shy away from nuclear power.

To criticism from the industry and from people who worry about nuclear safety is now added the criticism of the Kemeny Commission. The NRC, probably fearing that it may be about to go the way of the AEC, has responded to the criticism. Late in 1979, the commissioners announced that they would begin to enforce new, more rigorous safety and inspection programs. Among other things, they said they would require new safety features at each operating plant. The cost per plant was expected to amount to $10 to $50 million, depending on the plant's age and location. The commissioners also increased the size of their inspection staff. They planned to ask the utilities to devise emergency plans for the orderly evacuation of all people living within 10 miles of a nuclear plant. The NRC is also increasing the fines nuclear plants must pay for unreported safety violations. If a utility reports its own violation, however, the commission may reduce the fine by as much as 50 percent.

Are measures like these sufficient to meet the criticisms leveled at the NRC? Will they mean safer plants and more reliable personnel? Will they allow quicker licensing without sacrificing plant safety?

"The NRC program is superficial, inadequate, and does not attack the fundamental problems," is the answer given by Robert Pollard, who used to be on the NRC staff. Since leaving the NRC in 1976, Pollard has worked for the Union of Concerned Scientists. "The only things being fixed are the particular matters that went wrong at Three Mile Island, not the broad range of safety problems the Commission knows exist," he continued.

If Pollard is right, if the NRC is aware of problems it has no intention of doing anything about, it will be business as usual for the nuclear industry and its federal regulators. Only if there's a real change of attitude in the NRC will reform accomplish anything at all.

Similarly, the recommendation of the Kemeny Commission, replacing the NRC with a single director of nuclear regulation, would achieve nothing unless that director were prepared to look critically at the nuclear industry. Federal regulators—whoever they are—will have to begin putting safety first, ahead of nuclear promotion, or we will never have nuclear plant safety.

Even if reactors are made safe, problems will remain for nuclear power. There's the problem of whether or not to go into reprocessing. There's the problem of guarding fissionable materials against terrorists and madmen, and keeping nuclear weapons from getting into the hands of more and more countries. There's the problem of shipping and storing radioactive wastes. There's the problem of the deadliness of the breeder reactor—a grim fact we will have to learn to live with if we are to have a nuclear future.

Men and women in the nuclear industry have no doubt that such problems are solvable, although many people outside the industry disagree. One thing is certain, however: The problems exist, and if nuclear advocates hope to solve them, they must stop settling for easy answers to hard questions and put aside their comfortable assumption that when solutions are needed they will appear as if by magic.

9
PAYING THE NUCLEAR BILL

In the early 1950s, the infant American nuclear industry faced a problem: money. Back then, it looked as if finances might be a stumbling block on the road to the development of commercial nuclear power. The nation's electric utility companies, the companies that would own and run any nuclear power plants, were resisting the federal push for electricity from fission.

"Look," company executives said, "you want us to invest in nuclear power—to build huge new plants, to buy expensively processed uranium, and to 'burn' that uranium in complex nuclear reactors. But nuclear energy has never been tried on a commercial basis. How do we know it will work? You're asking us to take a big risk. After all, utilities are private, profit-making businesses. We are owned by the thousands of men and women who have bought shares of our stock. The stockholders, having invested their money with us, expect to get a profit in return. If they don't, they will sell their stock in our companies and put their money elsewhere. We will go bankrupt. Who will provide electricity for homes, schools, churches, stores, offices, and factories?"

To people in government, the message was plain. No guaranteed nuclear profits, no nuclear power. So Congress, the AEC, and other federal and state agencies took what action they could to turn the new technology into a sure thing financially. According to a 1980 estimate from the Department of Energy, the federal government has spent $37 billion to support the nation's nuclear industry so far. Nuclear critics charge that the actual figure is probably higher than that; government aid to the industry comes in so many forms that it is difficult to assess the total government investment.

Private Profit, Public Funding

First the government made research facilities available to private industry. The companies that might be persuaded to manufacture reactors were worried about investing in an untried technology. Now they could test that technology at various government laboratories. Some of these labs have already been mentioned—the Oak Ridge National Laboratory in Tennessee and the National Reactor Testing Station in Idaho, for example. Since the labs belong to the federal government, they are supported by government money. That reduces the financial risks of nuclear research to private industry.

The federal government offered other research aids. In some cases, the government has given outright grants of money to individual companies to pay for nuclear experimentation. In others, the Defense Department has shared its nuclear know-how with private industry. The military's information, of course, has been gained through research paid for out of public funds.

In addition, public money finances one entire step in the nuclear fuel cycle. That step is enrichment. So far, no private company has been willing to undertake this complicated and expensive job. All three of the uranium-enrichment plants in the United States are staffed and run by private business, but they are owned by the federal government. The fuel the plants produce is sold at cost to the utilities that run the nation's nuclear power plants. "At cost" means utilities pay only what it costs to process the uranium; the government makes no profit.

The government helps out at other steps in the fuel cycle. Congress recently voted to spend about $225 million to clean up the wastes abandoned at the West Valley, New York, reprocessing plant a decade ago. After the mill-tailings spill into New Mexico's Rio Puerco in 1979, government health agencies paid for testing local residents for excess radiation. At Three Mile Island, Met Ed officials expressed the hope that the federal government would foot a good part of their cleanup bill, which is currently expected to come to at least $1.5 billion.

Met Ed's hopes may be fulfilled. It is certain that government assistance to the nuclear industry must continue into the future—if nuclear power is to have a future. Reprocessing will probably have to be done at government facilities, both because it is expensive and because of the dangers of loss or theft of uranium and plutonium. (As we have seen, however, government plants have not been immune to losses of fissionable materials.) Storage of high-level wastes will almost certainly be left up to the federal gvernment. So add reprocessing—if it comes—and waste storage—when and if the technology develops—to what the government is already doing to subsidize nuclear power.

Nuclear Costs and the Consumer

Of course, it isn't "the government" that's actually paying the nuclear bill. It's the public. All those "federal monies" and "public funds" and "government grants" come right out of Americans' pocketbooks. It's the taxpayer who bears the hidden costs of nuclear power.

Taxpayers are consumers, too, and it is as consumers that Americans pay nuclear costs in other ways. The rates that consumers pay for the electricity they use are set by the Public Utilities Commission (PUC) in each state. And their electric bills have risen partly because of what the PUCs do to make sure that nuclear power is profitable for the utilities that invest in it.

The nation's PUCs have helped boost nuclear power by allowing nuclear plant owners to pass certain extra costs along to consumers. For instance, utilities have traditionally operated as a "cost-plus" industry. Cost-plus is business shorthand for buildings, equipment, and maintenance costs plus profit margin. It has meant that when a power plant begins to sell electricity, the consumers who buy it are charged for whatever it cost to build and equip the plant, plus enough extra to make a profit of a certain percentage.

How could this benefit nuclear power over other means of generating electricity? It costs more to build and equip a nuclear plant than one that runs on water or fossil fuel. In 1975, the Investor Responsibility Research Center, a Washington, D.C., group that analyzes investment possibilities, reported it would cost $638.4 million to build a coal-burning plant, compared to $811.1 million for a nuclear plant able to produce the same amount of electricity.

Suppose a $638.4 million coal plant and an $811.1 million nuclear plant were built side-by-side and the state PUC said that each might charge its customers enough to make an 8 percent profit. In the first year the two plants operate, stockholders in the nuclear facility stand to earn $38 million more than those who bought into the coal plant.

For years, cost-plus has made building an expensive nuclear plant just as attractive to the country's utility companies as building a lower-priced plant would have been. It has insulated investors against economic reality, and has passed the enormous costs of nuclear plant construction on to consumers.

Now the attitudes of PUCs toward passing on costs of nuclear power facilities may be beginning to change. For instance, after the TMI accident, Pennsylvania's PUC said that certain costs of that accident could not, for the time being at least, be handed on to ratepayers. But the nation's PUCs do continue to pass many other nuclear power costs on. PUC rules state that when a plant is forced to shut down, its owners must buy power from another utility to replace what the plant would normally produce. That way, customers will go right on getting the power they need. But PUCs add that the cost of buying replacement power may be passed on to consumers in the form of higher electric bills.

Again, the rule is identical for nuclear and nonnuclear plants. Every power station, no matter what its fuel, has to shut down occasionally for routine maintenance. When that happens, its customers pay the added costs of replacement power.

But nuclear plants are out of operation more of the time than fossil fuel or hydroelectric plants. Nuclear plants have a yearly refueling shutdown, and they may also have to shut down to allow workers to perform simple chores in highly radioactive areas. Nuclear plants are more likely than other plants to be closed so that new safety devices can be installed, or because of minor equipment malfunctions. They may also have to close because of safety violations. Overall, nuclear plants provide less than 60 percent of the power they theoretically could produce. Other kinds of plants produce over 70 percent of their theoretical capacity.

However often a nuclear plant is out of operation, and whatever the reason for the shutdown, it's the consumer who pays the bill. For instance, it took the Tennessee Valley Authority a year and a half to repair the damage caused by the Brown's Ferry fire. During that time, the plant did not operate, and the TVA had to purchase replacement power. The price of that

Refueling is a yearly necessity at nuclear plants. Workers watch as the reactor head is lifted off the core to expose the rods.

power—$18 million a month for eighteen months—was tacked on to TVA customers' bills. Metropolitan Edison customers will be expected to pay at least part of the cost of the TMI accident. In fact, one Met Ed official suggested that *every* American electricity user be required to contribute directly to the cost of the cleanup through a monthly utility-bill surcharge. A later recommendation, this one from a federal study commission, suggested that the government raise the cleanup money by imposing a new tax on all citizens. In New York state, Consolidated Edison customers' bills rose 5 percent due to the shutdown following the accidental spillage at the company's Indian Point plant.

Many Americans are beginning to wonder whether PUCs ought to be allowed to pass such costs on to consumers. Was it right to ask citizens of the Tennessee Valley to pay for design errors at Brown's Ferry? Is it fair to expect the public to pay for a nuclear plant's poor safety record? More to the point, is it *safe?* Would nuclear plants be better planned and maintained if the utilities—and not the consumers—had to shoulder the costs of shoddy design and upkeep?

Would nuclear plants also be safer if the United States tax system were different? Under federal law, a nuclear plant (like any other type of commercial enterprise) can assume that its buildings and equipment are wearing out by a certain amount each year. As plant and equipment age, they depreciate—become less valuable—and so less tax need be paid on them. That's reasonable.

What many people think isn't reasonable is that tax laws passed by Congress allow a utility to claim a full year's depreciation on a power plant even if that plant has been idle for 364 days out of 365. Operating a plant just a single day each year can save millions of dollars in federal income taxes.

That potential saving has close links to nuclear safety—and nuclear accidents. During the last two months of 1978, Metropolitan Edison employees repeatedly tested the new TMI-2 reactor. The results were not encouraging. Equipment failed over and over, and human errors abounded. But Met Ed officials knew that if they could only get the plant on line before January 1, they would save a bundle.

So the officials closed their eyes to the warning signs. TMI-2 went into operation on December 30, 1978, and as its owners later informed their stockholders, helped save $5 million in taxes by so doing. Less than four months later, TMI-2 was on the brink of a meltdown.

Open-handed as Congress has been to nuclear power in taxation policy, it has been even more generous in the matter of accident insurance. Without a Congressional measure known as the Price-Anderson Act, there might well be no commercial nuclear power in the United States today.

The Price of Price-Anderson

Melvin Price is a Democrat from Illinois. John B. Anderson is a Republican from the same state. Both have served for years in the House of Representatives. In 1957, just three years after the AEC gave nuclear energy the official go-ahead, Price and Anderson teamed up to give the country's utilities something they were demanding in return for investing in nuclear power. This was the assurance that if the new technology turned out to be less reliable than the AEC claimed, the utilities would not lose money by it. The two congressmen drew up a bill that says if a nuclear accident occurs, all its victims put together can receive from utilities no more than a total of $560 million in damages.

Why $560 million? As Congressman Price says, the figure came "out of thin air." It may have come from thin air, but nuclear proponents were certainly around helping it, or a similarly low sum, to emerge. Such a small amount of money would not have begun to pay back the people harmed in a nuclear accident even as long ago as 1957. Today, with inflation raising prices weekly, $560 million amounts to still less. If the Three Mile Island area had had to be evacuated, says one government insurance administrator, costs to local residents would have been anywhere from $3 billion to $7 billion. A $560 million limit would have meant virtually no recompense to those residents at all. And this estimate just takes evacuation into account. A meltdown would really have sent damages soaring.

But high as they might have gone, there would have been no headaches at Met Ed. The utility was protected by Price-Anderson. In fact, two years after the accident, the plant's owners and builders agreed to pay just $25 million in insurance settlements for financial losses to thousands of local businesses and individuals.

Among those who have criticized the Price-Anderson Act is Congressman Ted Weiss, a New York Democrat. Says Weiss, "Price-Anderson says that neither the designers, developers, builders, suppliers, technicians nor managers of a nuclear plant can be held liable no matter what the cause of an accident might prove to be. . . . It gives nuclear energy a competitive advantage . . . it encourages irresponsibility at each stage of plant development and operation; and it would deny victims of a nuclear accident just compensation for their losses."

Another critic, consumer lobbyist Ralph Nader, pounds away at one question: If nuclear reactors are safe, why does the industry insist upon the protection of the Price-Anderson Act? Either the plants are safe and should be fully insured, Nader says, or they are not safe, and the industry knows it.

The Price-Anderson Act, the tax structure, and government subsidies all provide financial assistance to the nuclear industry. The Price-Anderson Act and the country's tax structure mean that nuclear plant owners can be protected from paying the price of unsafe plant conditions. Who will pay? The public. We, the people, must pay the bill for the direct help the government gives the nuclear industry—and we stand to pay the price, in illness, death, and property loss, if the Price-Anderson Act and the tax laws help bring about more nuclear accidents in the future.

Does the public know how much of the nuclear bill it is paying now—and may be expected to pay in the years to come?

10
THE BATTLE OF WORDS

"Plutonium is one of the most carcinogenic [cancer-causing] substances known. . . . One pound, if uniformly distributed, could hypothetically induce lung cancer in every person on earth."

"But plutonium is not particularly *hazardous:* it cannot harm people it does not reach."

Two statements, two points of view. Together, they illustrate how difficult it is to get accurate, reliable information about nuclear energy and radiation. The first statement appears in a book by Helen Caldicott, an Australian pediatrician now living and practicing in the United States. Dr. Caldicott has opposed commercial nuclear power and nuclear weapons testing here and in her native country. Her book, *Nuclear Madness:*

What You Can Do!, emphasizes the deadly genetic and carcinogenic effects of ionizing radiation.

The second statement comes from a pamphlet put out by the Atomic Industrial Forum. The AIF's pronuclear position is evident throughout the publication. It contrasts plutonium's "immense benefits" to its "potential" risk, and makes that risk appear to be very, very remote. Only in "significant" quantity does plutonium cause lung cancer, it claims, and then only "many years later." And, it continues, "Safeguards have proved effective in stringently isolating plutonium from the public. . . . The risks of a commercial nuclear industry utilizing man-made plutonium are minimal."

Dr. Caldicott, on the other hand, is dismayed by what she sees as the perils of plutonium. Plutonium, she says, is readily absorbed by the human body. Breathed in, it causes lung cancer. Let into drinking water, it gets into bone marrow and the liver, bringing about bone and liver cancer. It can damage the iron-carrying red blood cells. Unborn children are particularly susceptible to plutonium, says Dr. Caldicott. According to her book, plutonium has escaped from civilian and military nuclear installations, and is already poisoning the world's population.

If Dr. Caldicott and the antinuclear activists who agree with her are right, we ought to shut down all nuclear power plants immediately.

If the AIF and other pronuclear groups are correct, commercial nuclear power and the breeder reactor appear to be good, and safe, answers to the nation's energy crisis.

The problem is deciding who is right. Unless we know what the facts are, we cannot make a wise decision.

We can get facts from both sides in the nuclear power debate, of course. But unfortunately, facts and facts alone are neither side's main concern. Each side is more anxious to tell us those facts that support its own point of view. And each side strives to present its facts in a way that will persuade us to share that point of view. The result is thousands and thousands

There's no lack of information—pro or con—about nuclear power. But is it all honest and accurate?

of facts, all skillfully marshaled into arguments in favor of nuclear power or against it.

A War of Words

Each side's viewpoint is made clear by its choice of words. Antinukers rarely speak of plutonium without describing it as "deadly," "toxic," or "lethal." When they mention that a radioactive element becomes harmless after 10 or 20 half-lives, they may preface "harmless" with "relatively" or "more or less." That suggests "harmless" isn't actually harmless at all. The possibility of a nuclear accident is likely to be discussed in terms of a "disaster," a "catastrophe," or a "holocaust."

Nuclear advocates have their own special vocabulary. Theirs is larger and more elaborate than the antinukers', because they have been using it longer. Since many pronukers are scientists and engineers, their lexicon is rich in technical jargon. It is designed to convince us that nuclear power is perfectly safe.

For instance, there are no accidents at nuclear plants. There are "transients" or "incidents." It was an incident that occurred at TMI. Other phrases used to describe the events in Middletown include "abnormal evolution" and "normal aberration." Was there a hydrogen explosion within TMI-2's containment structure? Not really. There was an "energetic disassembly." No fire followed the energetic disassembly; there was, however, some "rapid oxidation." In the end, the reactor vessel was contaminated by plutonium. Or rather, plutonium "took up residence" there.

Technical talk can be even more complex. The explosion that killed three plant workers at the Idaho National Engineering Laboratory in 1961 was a "power excursion." To avoid acknowledging that a breeder could experience a nuclear explosion, scientists refer to the possibility of "rapid reassembly of the fuel into a supercritical configuration and a destructive nuclear excursion." The same possibility can also be expressed as a "compaction of the fuel into a more reactive configuration resulting in a disruptive energy release."

Such circumlocutions are amusing, but nuclear advocates are as capable as anyone else of speaking in ordinary, forthright language when they want to. Sometimes the words they use are pretty unpleasant. In an article in the June 1979 issue of *Commentary* magazine, Samuel McCracken, assistant to the president of Boston University, offers his opinions of an antinuclear movie, *The China Syndrome,* about a near-accident at a fictitious California nuclear plant. McCracken directs harsh criticism at the movie, the people associated with it, and other antinukers. "Mendacity," "vicious fraud," "lies," "bamboozle," "a base desire for money," and "gross irresponsibility," are a few of the words and phrases he uses. Other nuclear proponents echo McCracken. Nuclear engineer Robert W. Deutsch, in a booklet entitled *Nuclear Power,* characterizes antinuclear demonstrators as "a nondescript group of people." By implication, he accuses nuclear opponents of being idle, fearful and timid, bullying, and against the American Way of Life.

Deutsch and McCracken are not the only name-callers among nuclear advocates. And nuclear advocates are not alone in casting epithets. The antinuclear Swiss historian and writer Robert Jungk has written a book called *The New Tyranny,* in which he describes nuclear advocates as "capable of anything," including criminal activity, if it will help them foist their technology upon the world. In Jungk's view, those who favor nuclear power are greedy, manipulative, and power-hungry. Those who oppose it are kind and loving. Such an assertion is as simple-minded as the notion that only the weak and shiftless speak out against nuclear power.

Such, however, are the assumptions that underlie much of the nuclear power debate in the United States today. Each side in the debate backs up its arguments with a carefully chosen vocabulary. Word by word, each side constructs propaganda aimed at getting the public to accept its version of the facts.

Advertising Messages

Much of the pronuclear propaganda reaches the public through advertising. The industry has spent millions of dollars writing and designing ads. Millions more have gone to buy ad space in magazines and newspapers and air time on radio and TV.

The ads, most of them at least, contain facts—but facts that are cleverly selected and presented. Their overall effect is to make nuclear energy appear to be safer, cleaner, and better in every way than any other means of generating electricity.

Here's a fact from a newspaper ad for the Mobil Corporation: "There is, for example, less radioactivity from nuclear power plants than from coal-fired plants, because coal plants release radioactivity from uranium and thorium when the coal is burned." True enough. But the radioactivity from a nuclear power plant is only *part* of the radioactivity released throughout the nuclear fuel cycle. Radiation can escape during uranium mining and milling. Radioactive wastes result from uranium enrichment. The radwastes at the back end of the cycle are an unsolved problem. In addition, there's always the possibility that a reactor accident could result in a massive outpouring of radioactivity. That couldn't happen at a coal plant. The Mobil ad takes an undisputed fact and twists it into an inaccurate comparison between nuclear plants and coal-fired ones.

Other industry ads hammer away at the theme that nuclear power is the cheapest source of electrical energy available today. The ads quote facts: The price of oil is soaring. Coal is getting more expensive to mine. In contrast, the price a utility company pays for uranium is *not* rising particularly fast. That means lower electric bills for consumers of nuclear power.

What ads like these gloss over is that there are hidden costs in the use of nuclear power, which consumers pay for. Our tax dollars support nuclear research. Tax revenues also pay the rising costs of uranium enrichment. Someday soon we may be paying for reprocessing and waste storage. The cost of generating electricity with U-235 is rising, just as the cost of generating electricity with coal and oil is. But fossil-fuel price increases are taking place right out in the open; the rising cost of producing electricity by fissioning U-235 tends to be disguised.

Hidden Persuasion

If pronuclear advertising is deceptive, so too is the other part of the industry's propaganda campaign—public relations. Public relations campaigns can be more misleading than paid ads are, because people are likelier to distrust ads and to question their claims than they are to be suspicious of a well-conducted P.R. effort.

Every nuclear power plant, every utility that invests in nuclear power, every pronuclear organization and trade association has a public relations staff whose job is to promote nuclear development. From these staffers come the glossy pamphlets and booklets, the newsletters and news kits, the fact sheets and press releases, the summaries and analyses that have been pushing commerical nuclear power for nearly thirty years.

At first glance, the P.R. looks convincing. Take this passage from a booklet prepared by the Edison Electric Institute, a nuclear trade organization. The booklet is called *Nuclear Power: Answers to Your Questions.*

> What is nuclear radiation?
> Man has an excellent scientific understanding of radiation. He should. He's been exposed to radiation since the beginning of time.

Aha! the reader thinks. Radiation isn't mysterious or scary at all. We understand it completely, it's always been around, and if it hasn't hurt us so far, a little more isn't going to do anyone any harm.

But wait a minute. "An excellent scientific understanding"? A hundred years ago, scientists had never even heard of radiation. Only in the 1930s were they realizing that radioactivity can be deadly. Today there is bitter disagreement among scientists over the hazards of low-level radiation. That "excellent scientific understanding" simply does not exist.

And what about, "He's been exposed to radiation from the beginning of time"? So he has. So what? Most experts agree that background radiation has always been responsible for a certain percentage of cancers and genetic changes in the population. They do not doubt that raising the dosage will result in more disease and abnormalities.

In the same Edison Electric Institute booklet, you will find this statement: "The Price-Anderson Act protects the public against the financial consequences of a nuclear accident. . . . Anyone filing a claim has only to demonstrate the amount of damages and that they resulted from the nuclear incident to receive compensation."

Well, demonstrating the amount of damages shouldn't be too difficult. If your home and possessions become so contaminated that you can never go back to them, you can figure that you are owed a hefty sum. Proving that the contamination resulted from an accident at a nuclear plant across town shouldn't be too hard, either. Collecting the money, though, may be a stickler. Under Price-Anderson, only $560 million in insurance funds will be available to pay all the victims of the accident. You will be lucky to receive a few pennies of every dollar that's owed to you. Furthermore, most home insurance policies contain a clause exempting coverage for nuclear accidents.

Another leading source of pronuclear public relations is the Atomic Industrial Forum. The AIF's statement that plutonium is not particularly *hazardous* because it cannot harm people it does not reach is typical P.R. *Of course* plutonium won't harm you if it doesn't reach you. Neither will a speeding bullet or a dose of arsenic. Nothing can hurt you unless it reaches you. How could it? But at least you can see a gun and taste poison. You are warned that danger threatens. Plutonium's alpha radiation gives no warning. In the same way, the industry's assertion that plutonium is harmful only in "significant" quantity has a reassuring ring. Reassuring, that is, until you realize that "significant" is meaningless in this case. By definition, a "harmful" amount has to be a "significant" one.

Dr. Helen Caldicott's allegations about plutonium, on the other hand, are not a bit reassuring. But what does the fact that one pound (.45 kg) of the substance could give lung cancer to every person on earth really amount to? No human being, no matter how brilliant or industrious, could devise a plan for distributing a pound of plutonium so precisely. Dr. Caldicott's statement tells us that plutonium is very toxic, period.

This statement, and Dr. Caldicott's book, *Nuclear Madness,* are good examples of antinuclear propaganda. So is Ralph Nader's *The Menace of Atomic Energy.* The books' titles betray their bias. Although both contain many accurate, provable facts, both use those facts in order to demonstrate that nuclear power is unsafe, uneconomic, and unnecessary.

Another example of antinuclear propaganda is the movie thriller *The China Syndrome,* which starred Jane Fonda and Jack Lemmon. That movie probably turned more people against nuclear power than a dozen carefully written and completely factual books could have done. Much of what *The China Syndrome* has to say about nuclear plants is accurate enough, although nuclear experts find fault with some of the technical detail. But technical accuracy is not the point of the film. *The China Syndrome* offers viewers a chilling picture of greed and lawlessness among nuclear plant owners and operators. It is powerful antinuclear propaganda.

Compared to pronuclear propaganda, however, antinuclear material is scarce. That is partly due to the fact that the antinuclear movement is relatively young. Before TMI, only a few people and organizations were concerned enough about nuclear power to speak out publicly. Pronuclear groups have been conducting P.R. campaigns for years.

Another reason for the relative scarcity of antinuclear propaganda is that the new movement is not tightly organized and has little money. Most antinuclear groups consist of people who get together from time to time to picket a nuclear power station or to rally to demand an end to new plant construction. Antinukers generally think they are doing well if they mange to print up an occasional newsletter to distribute to interested citizens.

Pronuclear groups do not have such money problems. The Edison Electric Institute's total 1979 budget was $14.6 million. And privately financed pronuclear publicity is only the start.

11 MISDIRECTION, MISINFORMATION, AND OFFICIAL DECEPTION

A good part of the pronuclear propaganda we see and hear comes from government officials and nuclear regulators. Some of it comes from the nation's most prominent scientists. It comes from members of Congress and from presidents. Over the years, hundreds of highly respected men and women have helped promote nuclear power by deliberately exaggerating its advantages and ignoring its dangers and unsolved problems.

One who apparently did so was President Dwight D. Eisenhower. In the 1950s, the Army was testing nuclear weapons in the Nevada desert. When ranchers there complained to the AEC that radioactivity was making their sheep fall sick and endangering their own and their families' health, the commissioners asked President Eisenhower what to do. Should they order an end to the tests? No, the President answered. Go ahead, but play down the radiation hazards. "Keep them [the public] confused," a former AEC chairman quotes Eisenhower as saying. The AEC did, assuring Nevadans that the weapons testing had nothing to do with their animals' illness. Today we know without a doubt that the illness and the testing *were* linked.

Confusing the public has been an important tactic in government and industry's campaign to develop nuclear power. In 1966, the AEC decided to keep secret the fact that more than 200 pounds of highly enriched uranium had disappeared from a Pennsylvania plant. The bomb-grade uranium has never been found.

More recently, a team of medical men headed by Dr. Thomas Najarian tried to investigate rumors that low-level radiation was causing sickness and death among Navy personnel at the Portsmouth Naval Shipyard in New Hampshire. After a two-year study, the doctors were ready to tell Congress about their findings—that radiation was indeed harming the Portsmouth workers. But according to Dr. Najarian, certain Navy personnel and government officials combined to discredit their report before they themselves could present it to Congress and the nation.

Low-Level Controversy

The Portsmouth dispute and the Nevada deception involve the military use of nuclear power, but each has implications for commercial nuclear power. The question of whether or not low-level radiation is really dangerous, and if so, how dangerous, is central to the industry's future. If low-level radiation turns out not to be a serious threat, the American public may well decide that going ahead with nuclear power is worth the risk of a few added cancers and genetic defects. But if it appears that low-level radiation jeopardizes the health of millions, as Dr. Caldicott and others believe it does, then nuclear power may be doomed.

Who's right about low-level radiation? In 1965, the AEC asked Dr. Thomas F. Mancuso of the University of Pittsburgh to begin a health study of workers at federal government labs in Oak Ridge, Tennessee, and Richland, Washington. The AEC told Mancuso that it wanted to find out whether people working with radioactive materials in the two labs showed a higher-than-normal incidence of cancer.

Nine years later, the study was still going on. One reason why it was taking so long is that radiation-induced cancers may not show up until thirty or more years after exposure. However, Mancuso was able to publish a few preliminary results. They showed that the Oak Ridge and Richland workers did have abnormally high cancer rates.

On December 2, 1974, the federal government's Energy Research and Development Administration began taking steps to remove Dr. Mancuso as head of the radiation study. (ERDA had by then assumed the AEC's job of nuclear promotion.) ERDA announced its intention of transferring the study to two separate research stations, one located in Oak Ridge and the other in Richland. In other words, the two centers would be studying themselves.

That prompted some members of Congress to launch an investigation. The findings of the House Committee are fascinating.

According to ERDA (by now part of the Department of Energy, DOE), it had decided to take the study away from the sixty-four-year-old Mancuso because he would soon be retiring. The committee learned, however, that in 1974 Mancuso actually had six years to go to retirement.

ERDA also claimed that a report drawn up by Mancuso's co-workers severely criticized his study methods. In response, one committee member whipped out a copy of Mancuso's colleagues' entire report. It included the flat statement that Mancuso's work was good. The ERDA witness had neglected to read that part of the report aloud to the committee.

The committee also learned that ERDA had spent months looking around in secret for someone to take over Mancuso's job. "Overtures to possible [replacements for Mancuso] must be carried out now in a clandestine atmosphere," said one ERDA memorandum.

Why the secrecy? If Mancuso's work were faulty, it would have been easy enough to fire him openly. If he had really been close to retiring, he would have stepped down automatically. The committee's conclusion: the AEC, ERDA, and DOE became anxious to get rid of Mancuso as soon as his work began demonstrating that low-level radiation does increase the risk of disease.

ERDA's resolve to hand the project over to research centers connected with the Richland and Oak Ridge labs bothered the committee, too. At best, they felt it presented a conflict of interest. At worst, it was a deliberate effort to manufacture results designed to reassure the American people that low-level radiation presents little or no hazard.

It may not seem likely that scientists at Oak Ridge or Richland, or anywhere else, would produce made-to-order results. But it can happen.

That doesn't necessarily mean that the scientists are dishonest. But scientists at reactor testing stations like Oak Ridge and Richland have close ties to nuclear energy and nuclear development. Like Glenn Seaborg and his fellow AEC commissioners, they understand nuclear reactions and radiation and are convinced that nuclear power is vital to this country and to its future. This conviction may lead them to overlook certain facts, or to ignore evidence that goes against their beliefs.

Radwastes and the NAS

For an example of how this can happen, consider the report on radioactive waste storage prepared by the National Academy of Sciences (NAS) at the request of the NRC. The 1979 report followed a two-and-a-half year study of the radwaste problem and concluded that high-level wastes can be stored safely. But it seriously questioned DOE's position that turning the wastes into a glassy substance and placing them deep underground is the best means of doing so. This vitreous storage method, the report stated, "may . . . be much less desirable than other solid waste forms."

The report was approved by a panel of NAS members, but it came in for rebuttal from DOE, then awaiting congressional approval for vitrification. Another critic was W. Kenneth Davis, a high executive in the Bechtel Corporation. Bechtel has been active in nuclear research for years.

Davis feared that critics of nuclear power would claim that the report showed no one was sure what to do with radwastes. The report's "conclusions and recommendations should be changed," Davis said, in the interests of "getting on with a realistic program which will enable nuclear power to move ahead."

So the report was changed. In its new form, it said that vitrification "may . . . be adequate." But, it went on, *"it cannot be recommended as the best choice."*

Davis didn't like the revision any better than the original. "Stupid statement," he scribbled on one page. On another, "I really object. Pure B.S."

The report was never published. Accepting the criticisms of Davis and DOE, NAS president Philip Handler denounced it as "flawed."

Cover-ups

Cover-ups are not uncommon within the nuclear establishment. One of them, brought about a strange, if temporary, alliance between the United States and the Soviet Union. In the late 1950s, nuclear wastes buried in the Ural Mountains, several hundred miles east of Moscow, exploded, possibly because of careless storage. A Soviet scientist now living in England claims that thirty communities in a 386-square-mile (1,000-square-kilometer) L-shaped area near Kyshtym were destroyed and hundreds of people were killed. The Soviet government tried to keep news of the event from getting out of the country.

In this, the Kremlin got help from the CIA. American officials knew of the accident, but refused to discuss it publicly. If the facts had come out, they might have alarmed people and caused Americans to wonder whether nuclear technology is as safe as government and industry claim. In the belief that the effects of nuclear accidents are worldwide, opponents of nuclear power, including Ralph Nader, attempted to clarify the facts and publicize the accident. News accounts appeared in several newspapers and periodicals. But the general public knew little about the incident until 1980, when reporters on the CBS-TV program "60 Minutes" broadcast the story.

Official concealment of potentially damaging information also occurred during the preparation of the Rasmussen Report on reactor safety. In one instance, the AEC asked Rasmussen to delete a section suggesting that inspection programs fail to turn up every safety defect. The Rasmussen Report was supposed to reassure people, not give them something new to worry about. In a memo, one AEC expert alerted others to the "disadvantage" of allowing the section to appear. "The facts [in that section] may not support our pre-determined conclusions," he wrote. The section was omitted.

Omissions like this one are intended to keep the public fuzzy about the true dangers of nuclear power. But cover-ups and misleading information deceive more than the public. They deceive—and ultimately threaten—the entire nuclear industry.

The Deceivers Deceived

People in the nuclear industry have begun to believe their own propaganda. That increases the chances that the industry will make the kinds of mistakes that will eventually convince the public to take a stand against nuclear power.

Behind the accident at TMI was the fact that Met Ed officials believed their own propaganda. Nuclear plants are safe—so it was okay to ignore malfunctions and go on line prematurely to get a tax break. Nuclear plants are safe—so they can be run by ill-trained operators and maintained in a cursory fashion. Just before the accident, a local newspaper questioned plant safety at TMI. Met Ed president Walter Creitz said the newspaper article was like "someone yelling fire in a crowded theater." Two days later, the "theater" was "on fire" for real.

Even then, Met Ed executives were so sure accidents could not happen that some of them managed not to believe what they were seeing. Right there in the control room, watching the temperature and coolant-level gauges, they did not seem to understand that a loss-of-coolant accident was taking place a few feet away.

After the accident, a TMI operator identified as "Instrument Man B" described the control-room scene to the NRC. About four hours into the accident, Instrument Man B said, he and four other workers were taking temperature readings at the reactor core. They found that temperatures were "anywhere from 690 to close to 4,000 degrees" (365.5 degrees Celsius to 2204 degrees Celsius). At 3,700 degrees Fahrenheit (2037 degrees Celsius) the fuel rods begin to suffer damage, and at 5,000 (2760 degrees Celsius), meltdown begins. Instrument Man B knew that. He continued "we did see one that I know for sure was 3,700, around in that area, and at the time both [a coworker] and myself turned around and looked at all three gentlemen [company executives] that were with us and said, 'This thing is melting down, that the core was uncovered.'"

Yet the gentlemen did not realize—or did not admit—that the core was exposed and meltdown a threat. One of the three explained that by saying there is a limit "to how much information it can reasonably be expected an official can assimilate in an emergency situation." Instrument Man B and his coworker, on the other hand, had no trouble taking in the fact that a core temperature of nearly 4,000 degrees meant danger.

Perhaps the company men really understood that, too. Perhaps they were just reluctant to let on how critical the situation was. It's certainly true that long after the accident became public knowledge, Met Ed officials were still denying that it was serious.

Company executives did not even inform government officials of how bad things were. As one federal investigator said later, "I am not sure whether the failure of the Metropolitan Edison guys was an attempt to downplay the seriousness of what was going on or just plain stupidity."

The NRC, too, complained about a lack of information from Met Ed. "We are operating almost totally in the blind," said one commissioner on the third day of the accident. "It's too little information too late," agreed an NRC staffer. By this time, the NRC was only beginning to piece together what was going on at the plant. "We are guessing [there] may have been a hydrogen explosion," an NRC official said as he scrutinized the sketchy reports from the plant.

Frustrated as they were by Met Ed's close-mouthed position, the commissioners were exceedingly reluctant to give out information themselves. They, like Met Ed's spokespersons, tried to keep outsiders from realizing just how threatening the situation was. Even the President couldn't get straight facts. In a phone call to the White House, the NRC said that keeping the core covered was no problem.

Pennsylvania governor Richard Thornburgh had difficulty getting information, too. Thornburgh wanted to know whether he should order an evacuation. The commissioners' responses were vague. "I wonder if, well, oughtn't we think about at least urging people who are real close in, they don't have to be around here now, to, if they've got relatives twenty miles away, to go visit them," mused one.

"Yeah, but don't you—if you're going to take that kind of a step—don't you have to be more direct about it?" another wondered. "I mean, you can't sort of—the agency to whom they would look for advice—you can't sort of toss it out and say well, you know, golly, maybe—"

It's not surprising that the governor complained about a lack of solid information.

Nuclear News and the Media

Members of the news media had gripes of their own. According to a special panel of Kemeny Commission members, they had reason to be unhappy. As the panel reported, Met Ed and the NRC decided at the outset "that bad news was not something that the public ought to hear." The two made a "conscious decision . . . to impart only available evidence and to avoid discussing its implications," the report continued. Panel members also spoke of "the utility and NRC's desire to minimize and forestall bad news." They called one of the NRC's press statements "a classic example of fudging."

Met Ed and the NRC could claim one sound reason for minimizing the accident's seriousness. They wanted to avoid panic among the public.

Are we to believe that was the only cause of the official reluctance to give out information? Surely Met Ed could have been candid with the NRC in private without alarming anyone. But we know that the nuclear industry and its regulators have a long history of cover-up and deception. They have concealed bad news when they could, and made bad news look like good when they could not. They have misrepresented facts and tried to lull the public into accepting nuclear power as virtually risk-free. Given that history, it is hard not to suspect that Met Ed and the

NRC were as anxious to protect themselves and their technology from criticism as to keep the public from panicking.

For whatever reason, news reporters did have trouble finding out what was going on at TMI. They besieged company executives, the NRC, and other experts with one question after another. Sometimes, the answers they got did not fit together.

On the third day of the accident, for instance, a CBS-TV reporter spoke with a Pennsylvania civil defense officer. "Nobody is in any danger," the worker said. The reporter then turned to Governor Thornburgh, who informed him that, "based on advice from the chairman of the NRC," he was suggesting that pregnant women and young children leave the area. Similar contradictions were apparent throughout the broadcast, and contradictions also cropped up on other networks and in newspaper and magazine reports.

When the accident was over and the reporters had gone home, many in the nuclear industry condemned their performance during the emergency. In his booklet *Nuclear Power,* Dr. Robert Deutsch accuses the news media of doing "everything in its power to indicate that we were about to witness the greatest tragedy of all time." Deutsch and others like him believe that reporters and editors are so eager to militate against nuclear power that they ignore the dangers the public faces from other sources, such as auto accidents, tornadoes, train derailments, spillage of poisonous chemicals, and so on.

It's true that during the time of the accident, the news media focused most of their attention on TMI. But that seems natural. Newsworthy events were going on there. It's also true that reporters used some alarming words and phrases about the accident. On March 30, CBS anchorman Walter Cronkite opened his broadcast by saying, "The world has never known a day quite like today." Cronkite also referred to "the horror tonight" and to the "specter" of a meltdown.

The trouble was that never before had reporters had to inform the public about such a situation. They had never had to try to make sense of nuclear truths, half-truths, lies, and contradictions during a nuclear emergency. Few of the reporters knew enough of the basic facts about nuclear science and technology to understand what was going on.

The nuclear establishment must share the blame for that. Its P.R. specialists have done their jobs all too well. They have plied reporters with information kits and press releases that contain facts—but not all the facts, or facts that are presented in misleading ways. They have invited reporters to tour nuclear plants, and have pointed out safety features—but neglected to mention that those features don't invariably work as planned. They take reporters to the "hot" side of a plant—and assure them that if they should get contaminated, they can "brush the contamination off, just like dirt." They have listened to reporters' questions about problems like radiation leaks and the difficulty of storing radwastes—and replied that such problems simply do not exist.

The nuclear industry has misinformed the press, as well as the public, and people in the industry should not be surprised when that misinformation comes back to haunt them.

Of course, members of the press are to blame, too. They ought to have started to work long ago to ferret out the truth.

Will they start now, in the wake of TMI? And will the industry begin to discuss the real drawbacks of nuclear power as well as its advantages?

It's certain that reporters are displaying more interest and more knowledge about nuclear power than ever before. The media are full of stories about safety defects at this or that plant, about the unresolved matter of radwastes, and about other nuclear questions.

But do these stories concern real problems? Or are they just a new form of antinuclear propaganda? Many nuclear advocates insist they are the latter.

At the same time, the nuclear industry is stepping up its effort to communicate with the public. Advertising and P.R. budgets for nuclear trade associations are rising. An entirely new pro-nuclear group, the Committee of Energy Awareness, was formed to encourage utilities and other corporations to establish "citizen groups" that will back nuclear power. Advises one committee manual, "The corporate participant should help provide the humor, ideas, linkages, resources and a sympathetic shoulder. . . . Corporate participation can help a citizen group multiply and influence decision makers."

It doesn't sound as though very much has changed in nuclear propaganda.

12
OUR ENERGY FUTURE

Should the United States push on with nuclear power? Or should we declare a moratorium and slow down the growth of the industry? Should we search for alternate energy sources so we can afford to give up commercial nuclear power altogether? There are questions all of us will be helping to answer in the months and years ahead.

Many people believe that nuclear power is so unpredictable and so hazardous that the only safe course is to shut down all the nation's nuclear plants immediately. Dr. Helen Caldicott shares this feeling. Others take a less extreme position, calling for an end to new plant construction and for a gradual phase-out of existing plants. The nation's growing antinuclear movement includes men and women with both points of view.

The Antinuclear Movement

Since the accident at Three Mile Island, over three-hundred organizations have joined forces to work for the eventual end to the use of nuclear energy for both civilian and military purposes. Operating as the Coalition for a Non-Nuclear World, organization members are finding ways to make their views felt at the voting booth and in the halls of Congress, in state and local government, and throughout the news media.

A few antinuclear activists have carried their fight to the very gates of some of the nation's nuclear power plants. A nuclear power station under construction at Seabrook, New Hampshire, has been the target of angry protesters whose aim is to seize control of the site and halt construction there. Several clashes between antinuclear forces and the police and National Guard have resulted in the demonstrators being beaten back by club-wielding law officers.

Violence is not common in the antinuclear movement, however, and most activists prefer to make their point in peaceful ways. Mass marches and demonstrations are a popular form of protest, especially among young people. About a month after TMI, 65,000 demonstrators gathered in Washington, D.C., to show their support for a nuclear moratorium. A few months later, nearly four times that number participated in a rally in lower Manhattan, carrying placards with antinuclear slogans and listening to songs and speeches about nuclear dangers. Thousands more turned out for similar demonstrations in a dozen cities across the United States. Additional marches and rallies followed.

Educating the public is a vital part of the nonviolent antinuclear movement. Teach-ins in some towns and cities have stressed nuclear hazards to concerned citizens. In other places, nuclear opponents have invited industry spokespeople to air their side of the issue as well. The opponents do not mind sharing a platform with nuclear proponents because they are positive that when Americans understand the facts about nuclear energy, they will decide to abandon it.

The antinuclear educational effort also includes direct mailings to hundreds of thousands of American homes. Activists work long hours drawing up lists of men and women who may be interested in learning more about nuclear energy. Others write and edit the pamphlets, newsletters, and appeals for financial contributions that are sent to people on the lists. Antinukers are also careful to keep close track of news reports about nuclear power, speaking out whenever they suspect that the press is downplaying its dangers.

Antinuclear activists are turning to direct political action, too. Some takes place at the local level. At their annual town meetings in the spring of 1980, Vermonters were scheduled to debate whether or not to permit mining of the uranium deposits that underlie the Green Mountains. Going ahead with exploratory mining might have meant new jobs and higher incomes in a relatively poor state, but antinuclear activists pointed out that it could also mean serious health problems. They reminded residents of the dangers faced by people who live near uranium mines and mills in several western states. By summertime, when the meetings were over, citizens in one town after another had said no to uranium mines. Elsewhere in the country, particularly where nuclear plants are operating or where nuclear wastes are stored, inhabitants are beginning to work for local laws to force nuclear industries to provide safer working and living conditions.

On a state government level, antinuclear activists have mounted campaigns to collect enough signatures to get nuclear referendum questions on the ballot. In 1976, Californians had a chance to vote on a proposition to impose strict safeguards on the state's nuclear industry and to waive the Price-Anderson Act there. Montana residents got to vote on keeping nuclear power out of their state altogether. In Maine, antinuclear activists began a drive to enable voters to decide whether or not they wanted to shut down the state's only nuclear plant. The activists needed 27,026 signatures to get the question on the ballot; they got over 55,000. In other states, too, nuclear opponents have succeeded in initiating similar nuclear referendum questions.

To their disappointment, the antinukers have had considerably less success in getting their propositions passed by the voters and made into law. The people of Montana did, in 1978, decide to ban the construction of nuclear plants in that state until adequate means of waste storage are found. And in 1980, in Washington and Oregon, two states that had previously defeated antinuclear questions, voters agreed to place some limits on radioactive waste dumping within state boundaries. But in Maine, where the referendum called for closing an operating plant, 60 percent of those who voted cast their ballots in favor of keeping the plant open. Interestingly, the vote to *close* the plant was heaviest among those Maine residents who lived closest to the facility.

Antinuclear questions have also lost in such states as California, Missouri, South Dakota, Arizona, and Ohio. Part of the reason for their failure is financial; whenever a nuclear referendum appears on a state ballot, pronuclear groups around the country rally to defeat it. Organizations like the Atomic Industrial Forum spend thousands of dollars on advertising aimed at persuading voters that nuclear power is not only safe, but that it is essential if Americans are to continue having good jobs, high pay, and affluent life-styles. In Maine, nuclear proponents outspent opponents by almost 4 to 1. In Ohio, the margin was 80 to 1, and in Arizona, 90 to 1.

Congress is another arena for antinuclear activity, and scores of Congressmen and women have found themselves bombarded with letters and information from nuclear opponents. The antinuclear movement can claim some converts, or partial converts, among members of Congress. Some who once strongly backed the nuclear industry sound a little less enthusiastic today. Antinuclear activists hope that Congressional sup-

At a polling place in rural Maine, a man signs a petition for an antinuclear referendum question. The petition drive was successful, but Maine residents voted to keep the state's nuclear power plant in operation.

port for nuclear power will diminish as the Senators and Representatives learn more about the industry's very real problems.

But getting Congress to vote against nuclear power is a tough assignment, and most antinuclear activists know it. Shortly after the TMI accident, both the Senate and the House of Representatives rejected a bill that would have imposed a six-month moratorium on new plant construction. The House vote was overwhelmingly pronuclear: 254 to 135.

Congress is not the only branch of government that favors continued nuclear development. Our presidents have been committed to nuclear energy, and the executive branch will probably continue to push for new nuclear plants to supply electricity to Americans. Ronald Reagan, elected president in 1980, has announced his active interest in building a commercial breeder reactor capable of generating enough power for a city of 200,000 by 1988.

The breeder, and nuclear power in general, also continue to receive strong support from people within the industry. They believe that the country needs nuclear power, and are certain that any problems the industry faces can easily be solved. These people, like the antinuclear activists on the other side, are using all their resources to convince us that their point of view is correct.

The Risks of Risk Analysis

One of their arguments concerns risk. Some nuclear proponents concede that nuclear power does perhaps present a measure of risk to society. But they maintain that other modern technologies can be risky, too, and point out that we seem perfectly willing to live with those risks.

It's risky to drive a car or ride in an airplane. It's risky to approach a dam that might collapse. It's risky to eat foods with certain additives or preservatives. It's risky to be anywhere in the vicinity of a railroad freight car that could jump the tracks, releasing a load of deadly gas. It's risky to live near a dumping ground for poisonous chemicals. We are surrounded by risks, and we manage to take them in our stride. Why, then, do people get so excited about the idea of nuclear risk?

The answer, say nuclear supporters, is that people don't understand what that risk really is. In an effort to educate the public, nuclear proponents have constructed elaborate formulas that allow a comparison of the risks of nuclear power with many of the other risks we face daily. Set down in charts, graphs, and tables, these comparisons appear to show that generating electrical power from nuclear reactions is extraordinarily safe.

For example, an analysis by Bernard L. Cohen and I-Sing Lee of the University of Pittsburgh shows that forty-eight separate factors are more likely to shorten life than is radiation from the nuclear industry. Among the life-threatening factors the two men studied are having less than an eighth-grade education, being murdered, using illicit drugs, and riding a bike. Most dangerous of all, they discovered, was being an unmarried male. That can take almost ten years off a normal life span. Radioactivity from the nuclear industry, however, deprives the population of only a 0.02 day of life.

Other risk analyses by nuclear advocates focus on the comparative dangers of various energy sources. Uniformly, the studies conclude that nuclear energy is safer than energy derived from fossil fuels, wind, water, or the sun. Herbert Inhaber, a Canadian physicist who serves on his country's equivalent of the NRC, did a survey that shows that solar power is twelve times riskier than nuclear power. Wind power, according to Inhaber, is a whopping seventy times riskier.

How accurate are these figures? Not accurate at all, according to many scientists. One problem is that Inhaber bases his figures on predictions that workers connected with solar power will spend 120 days off the job for every 10 days off for people associated with nuclear power. These predictions reflect the fact that technical failures are more likely at solar installations than at nuclear power plants. But Inhaber is comparing present-day nuclear plants with present-day solar technology. The two technologies cannot be equated. Billions of dollars have gone into nuclear research; very little into solar. Nuclear technology is far more advanced than solar technology, and no accurate conclusions can be drawn from comparing the two. The same is true of Inhaber's claim that wind power is seventy times riskier than nuclear. Wind-power technology is in the crawling stage, too. What the Inhaber analysis really says is that right now, wind and solar technologies are more subject to breakdown than nuclear technology is. It says nothing about the specific risks of nuclear power—risks that go far beyond mere equipment failure to partial or total meltdown, radiation releases, nuclear weapons proliferation, the lack of safe storage technology, and so on.

Risk analysts make other faulty assumptions. A pair of English scientists demonstrated that nuclear plant workers face only a slightly greater risk from radiation than does the population at large. What this analysis failed to consider is that the effects of ionizing radiation don't just go away after forty or fifty years. Genetic damage to workers may show up in their great-grandchildren or great-great-grandchildren. The damage may be delayed, but delay doesn't reduce it.

A more basic false assumption is that all risks can be measured in numbers. As human beings, we meet risks in real life, not in a computer model or a mathematical equation. Writing in the magazine *Technology Review,* managing editor Steven J. Marcus ponders what Inhaber would have made of TMI. (Inhaber's analysis was published before the accident.) "Equipment failures; 'human error'; misinformation; jurisdictional problems; carelessness; and simply not knowing what to do next." there's no way, Marcus writes, that anyone could turn those accident ingredients into a language of numbers. The risk cannot be examined mathematically.

A Nuclear World?

Another argument that nuclear proponents use to try to convince Americans that they ought to go ahead with nuclear development is that other countries are doing it. West Germany has ten nuclear plants, Sweden has six, and Canada, eight. Japan is reprocessing some of the spent fuel rods from its twenty nuclear power plants. Brazil, Argentina, and Mexico are in the process of constructing nuclear power reactors. England has found places to temporarily store wastes from its thirty-three plants. France, with fifteen plants already on line, is building more. Currently, the French are assembling the world's first commercial breeder. Built with the cooperation of West Germany, Italy, and England, the French breeder should be generating electricity sometime during the 1980s. By the year 2050, one American nuclear advocate says, three fourths of the world's energy will come from breeders. Worldwide, there will be five thousand reactors, producing nine times the total energy we generate today.

Other nations are forging ahead with nuclear power, nuclear proponents say, so why not the United States? If we hold back we will be left behind technologically. What's more, holding back won't answer the concerns of the antinuclear forces. Nuclear radiation does not respect national boundaries, and Americans could be threatened by a nuclear accident in Canada or Mexico almost as much as by an accident in the United States. Even routine radiation releases from plants in countries as far away as China or Japan could pose a danger as the winds of the upper atmosphere spread radioactivity around the world.

Antinuclear activists agree that radiation from one country can endanger people in all other countries. That's one reason why they call themselves the Coalition for a Non-Nuclear *World.* Their ultimate aim is an end to the use of nuclear energy, civilian and military, in every single nation. But the activists do not concur with the nuclear industry's claim that other nations will slip ahead of the United States if we declare a nuclear moratorium. On the contrary, they assert that *no* nation will ever conquer nuclear technology because its problems are simply unsolvable.

True, they say, France has plunged enthusiastically into the age of nuclear power. But protests about new plant construction are just beginning to be heard in Brittany, in western France, as people there start to worry about the dangers of accidents and resultant radiation leaks.

Yes, West Germany was long in the forefront of nuclear development. But the Germans are having second thoughts, and no new nuclear power stations have been ordered in Germany since 1975. Plans for a nuclear complex of fabrication plants, generating stations, fuel-rod storage pools, reprocessing facilities, and dumping grounds are running into strong public opposition.

Sweden does have six nuclear plants in operation, and in 1980 the Swedes voted to go ahead with plans to build six more. The vote reflects the fact that Sweden is desperate for energy. It has no coal or oil reserves of its own—but it does have the largest uranium deposits in Western Europe. The vote also demonstrates that the Swedes are deeply divided over the wisdom of taking the nuclear path. In their 1980 referendum, nearly 40 percent of the voters, including the Prime Minister, cast their ballots in favor of an immediate end to all nuclear power production. The antinuclear forces lost, and nuclear development will continue for the present in Sweden. However, the large antinuclear vote probably means that the country will have no more than twelve plants in all, and that when those twelve are decommissioned, no new ones will be built.

Japan did begin reprocessing not long after the TMI accident, as groups like the AIF are quick to point out. But nuclear industry officials are less eager to advertise the fact that Japan tried reprocessing before and ran into such difficulties that its one reprocessing plant was closed for fifteen months. Industry people are not any happier to admit that almost as soon as reprocessing got under way in Japan again, radioactive nitric acid was found leaking from a 7-ton tube, and the plant had to be closed a second time.

England is looking for permanent storage places for its radwastes—in Scotland. The Scots resent being made the recipients of toxic wastes produced to provide electricity for the English, and they are protesting. English scientists themselves are far from satisfied that their storage methods are safe, and they are actively searching for new, better technologies.

So it is not true to say, as nuclear advocates do, that all the technological problems have been solved. Nor is it true that the nuclear industries of other nations are developing ahead of ours. Neither of these arguments in favor of going on with nuclear power is very convincing.

More convincing—at least at first—is the American nuclear industry's argument that we must continue with nuclear power in order to meet the nation's future energy needs.

That our demand for energy will rise in the next decades is

accepted as fact by those who urge nuclear development. They point out that the country's population is increasing. More and more people are going to need more and more jobs, food, clothing, homes, recreation equipment, and so on. It's going to take a growing economy to supply those needs. A growing economy, nuclear supporters say, cannot exist without ever increasing amounts of energy to fuel it. Oil is costing more and getting scarcer; coal threatens the environment. That leaves nuclear energy as the best, in fact, the only possible, answer to our future energy needs.

Nuclear opponents do not accept this argument, either. They believe that we have many other potential energy resources, enough to fill all our demands without resorting to either coal or foreign oil.

Alternative Sources of Energy

There's water power, for instance. In our coastal areas, great tides ebb and flow, ebb and flow. Harnessing those tides could provide large amounts of power to generate electricity. Inland, thousands of abandoned dams lie idle. According to the U.S. Army Corps of Engineers, if those dams were only partially used, the United States could double its hydroelectric output.

Wind is another power source. The Department of Energy and the National Aeronautics and Space Administration are testing windmills of various sophisticated designs in several states and in Puerto Rico. Within twenty years, specialists say, wind-power alone could supply 4 percent of the country's electricity.

Electricity can also be generated by burning wood or other plant substances. Garbage and other refuse can be burned to produce steam to run electrical generators. The city of Burlington, Vermont, is producing electricity in commercial quantities in a wood-burning plant. In the Midwest, manure is being converted to energy-yielding gas. In Texas, guayule bushes produce a rubbery material that has certain resemblances to crude petroleum. If an area the size of Arizona were planted with guayule, claims Nobel Prize winner Melvin Calvin of the University of California, the fuel output would equal current United States oil imports.

Energy to drive our cars and to run many kinds of machinery might come from manmade, or synthetic, fuels. Coal can be processed to produce both oil and gas, for example, and shale to produce oil. Synthetic fuels can also be made from corn, sugar cane, and sugar beets. Most synthetic-fuel processing is too expensive to be practical today, but further research might make it economical.

Geothermal energy is yet another possibility. Deep inside the earth, where they are heated by red-hot molten lava, are hot-water pools. In California, the Pacific Gas and Electric Company is generating about 6 percent of its electricity from a hot-water geyser. That's enough to supply the power needs of a city the size of San Francisco. A scientist at DOE states that geothermal power is already competitive with fossil fuels.

Another warm-water energy source is available in tropical and semitropical ocean areas. Here the surface water has much more heat than the water farther down. A substance like ammonia gas can be piped from the cooler water through the warmer layer. This heats the ammonia sufficiently to turn it into steam to drive a turbine.

Solar power, too, is a possible alternative to nuclear power. Energy from the sun can be used either passively or actively. Passive solar energy means letting the sun in when it's needed and keeping it out when it's not. Large south-facing windows let in the sun's warming rays on a winter day. Thick curtains over the windows keep heat from escaping at night. In the summer, the windows are opened or shaded to allow cooling.

In active solar-energy use, solar collectors are placed on a rooftop or the side of a building. The collectors conduct heat energy into the building, where it can be stored, perhaps in water tanks, perhaps in small heat-absorbent rocks. When it is needed, the heat is slowly released and used. Like wind and water power and energy from organic matter or geothermal sources, solar power has come to be known as "soft" energy. Soft-energy resources have the great advantage of being renewable. Sun, wind, and water are inexhaustible; they cannot be used up the way coal and oil can. Even wood, which is used up as it burns, is renewable in a way that fossil fuels are not. Only a few years are needed to grow a tree compared to the millions of years it takes nature to form a ton of coal or a barrel of oil.

Besides being renewable, soft-energy sources are flexible. A home can be solar-heated either actively or passively—or both. It can have a small windmill in the backyard and a wood stove in the kitchen. Similar diversity can characterize soft-energy use in a school or factory.

But diversity and flexibility are not enough, protest the proponents of nuclear energy. The soft technologies will never produce enough energy to provide the basis for a growing economy. Wind, water, wood, and sun simply cannot power all the nation's factories and generating stations, its transportation systems, its farms and hospitals. America's energy needs are so great that only the hard-energy approach—producing electricity in complex, centralized generating plants—will fill them.

Hard vs. Soft Energy

The hard-energy path is the one that America has followed for decades. Producing electricity by burning coal or oil is a hard-energy technology. So is producing electricity by fissioning U-235. It's no coincidence that the nation's largest petroleum companies also control much of the nation's supply of uranium. The oil companies, the public utilities, and most of our largest industries are determined to keep America bound to hard technology.

Why? Because hard energy promises the biggest profits.

Once, Americans did rely on soft energies. Homes were heated with wood stoves or fireplaces. Houses usually faced south, and windows were placed so as to admit the greatest possible amount of sunlight. Waterwheels provided most of the energy needed to grind grain and to power early factories. There were no public utility companies, because there was no need to produce vast amounts of energy.

Now things have changed and, protected by laws that guarantee hefty profits, today's public utilities are producing and selling more than 2 billion kilowatt-hours of electricity a year. Suppose that changed again, and Americans reverted to soft technologies for many of their energy needs. A utility would have to stop charging for its services every time the wind happened to blow at more than 15 miles an hour. No meter reader is going to try to estimate how many sunbeams are absorbed through rooftop collectors. As the demand for their services went down, utility company profits would plunge.

That's why the utilities—and the other industries that have a financial stake in hard-energy technology—are less enthusiastic than antinuclear activists about pursuing research and development in solar power and other soft-energy possibilities. It's also why they publicize studies like Herbert Inhaber's comparative risk analysis—the one that purports to show that wind and solar technologies are riskier than nuclear energy.

Another tactic of those who favor the hard energies is to encourage the spending of government money on hard, rather than soft, energy research. Groups that favor nuclear energy and other hard technologies have spent huge sums of money on Congressional lobbying efforts.

Pressure can be applied indirectly, too; for example, by exaggerating the potential costs of alternative soft technologies. A 1980 report by the National Academy of Sciences predicted an expenditure of $3 *trillion* if solar energy is to provide as much as 20 percent of the nation's energy needs by the year 2000.

Challenged by one scientific group to document their $3 trillion figure, the people who prepared the report answered that it had been a "back of the envelope" calculation. Where is the "envelope" with the calculation? No one can find it, was the answer. The challenging group refigured the solar cost and learned that the NAS report had overestimated it by 250 percent. In addition, there were other substantial solar savings that the report ignored—apparently deliberately.

At the same time that they work to oppose soft-technology development, nuclear proponents are working to ensure that solar power, if it does come, comes in a form that is adaptable to a hard-technology system. Some companies are designing huge solar satellites that may one day be capable of circling the earth, collecting energy from the sun and sending it back to power stations where it will be used to generate electricity. Public utilities will have fewer objections to solar power if it can be made to fit into a centralized hard-technology system.

Future Fusion?

Another form of solar energy that intrigues hard technologists is fusion. It is the energy of fusion that produces the tremendous light and heat of the sun and other stars. Stars are giant "furnaces," in which the nuclei of hydrogen atoms (each with one proton) are forced together to form helium atoms (each with two protons). This forcing together, or fusing, is what gives fusion its name.

Someday, fusion reactions on earth may provide us with a new means of generating electricity. Fusion research has been going on for three decades in the United States, the Soviet Union, and other nations, and physicists at Princeton University in New Jersey think now that they are on the verge of demonstrating that controlled fusion is technically possible. If their experiments succeed, commercial electricity production at fusion plants might begin around the year 2000. The 1980 Magnetic Fusion Engineering Act makes that year the target date for opening a commercial demonstration fusion plant.

Like fission, fusion is a nuclear reaction, but it has several advantages over fission. Its "fuel" would be a form of hydrogen that can be extracted from water and lithium (a light metal) ore. Radioactivity would be produced, but in a form that has a half-life of only twelve days. Although the components of the reactor vessel would gradually become highly radioactive, their safe storage is expected to be an easy matter compared with safely storing the vast amounts of radioactive wastes now being produced by fission. Finally, since fusion does not involve uranium or plutonium, fusion would not contribute to nuclear weapons proliferation.

With all its possible advantages, though, fusion has one great disadvantage: the day when it might be used to help supply the nation's energy is far off. That disadvantage is not shared by the renewable soft technologies. Given adequate funding and a chance to compete financially with the hard technologies, power from the sun, wind, water, and other soft-energy sources could offer us a way out of the energy crisis. Even now, the soft technologies provide 6 percent of America's total energy. Nuclear power provdies only 4 percent.

There are other signs that the soft technologies are gaining in favor. A recent study shows that half of all New England homeowners use wood for part or all of their winter heat. A growing number of homes around the country display solar collectors, windmills, or other soft-energy equipment. Congress has passed a bill to spend up to $88 billion over a seven-year period to develop synthetic fuels. Such fuels, elements of hard technology, might fit well into a diversified system of soft and hard technologies supplementing each other. Most important of all, perhaps, people are starting to look carefully at reports like the Inhaber analysis and the NAS report on solar energy to see what they are *really* saying about soft-technology feasibility. What people learn, even from such negative reports, could well

Our Energy Future **149**

incline them to favor the soft-energy paths, at least in part.

There's still another energy source available to us, a source that has been called "our cheapest and cleanest." This "source" is conservation.

We Americans don't just use more energy than any other people, we waste more. Many of us drive our cars more than we have to, and we drive them at energy-wasting speeds. We set our thermostats higher than necessary in winter, and leave the air-conditioning on in empty houses in summer. We use lights and electrical appliances lavishly.

We can cut down our energy use, and it may not be as difficult as we imagine. Industries can find ways to use the excess heat energy that presently escapes into the air or into nearby bodies of water. So can we, in our homes.

Even the United States Army claims energy conservation victories. In 1979, the Army set a goal of an 8 percent reduction in energy consumption. Insulation was added to buildings, and heating and cooling levels changed. New training techniques were used. At year's end, the Army found its energy use had dropped by 9.5 percent.

Did that saving mean a slowdown in Army activities? According to the nuclear industry, using less energy will result in a sagging economy with fewer jobs and consumer goods available. When the Army cut its energy use, it must also have cut its training requirements, its use of machinery, and its building program.

In fact, it did not. In 1979, the Army continued mechanizing its forces, upped the number of its training missions, and went right on constructing new facilities. Conservation did not mean sacrifice for the Army, and it need not mean a terrific sacrifice for the country as a whole, either.

So, should the United States continue to increase its use of nuclear energy? Congress, Presidents, and the nuclear establishment urge us in that direction. Our needs demand it, they say, and the technology is there. But other forces push us the other way.

Americans have reduced their demand for oil and electricity. Many are now aware of the dangers of nuclear power. They have learned to look at nuclear advertising and P.R. with a more critical eye. They are asking their Congressmen and -women to reconsider their past support of unlimited nuclear research and development.

Construction continues on the Seabrook, New Hampshire, nuclear plants. But construction costs have risen dramatically—there and elsewhere. With all its unsolved problems, what kind of future is nuclear power likely to have?

Some business people, bankers, and utility owners are also beginning to think twice about nuclear power. Plant construction is getting more and more expensive, and it commonly costs 100 percent more to build a plant than was originally estimated. One plant now under construction on Long Island is expected to cost $2.2 billion before it is finished—eight times the figure quoted in 1969, when the plant was proposed. This makes investors wary of putting more money into nuclear power. Owners of plants under construction are having trouble coming up with the cash to finish the facilties and get them on line. New-plant orders are slowing down. A sort of moratorium is in effect despite the reluctance of Congress to call for one officially.

What about the future? Will we go on with nuclear power and find it means a world set free? Or will a nuclear world prove to be a world headed for grief?

The debate continues, and the decision—nuclear or not—is up to us. We must endeavor to decide wisely, to choose the best way for a world like our world, in a time like our time.

BIBLIOGRAPHY

Alternate Nuclear Fuel Cycles. American Nuclear Society, La Grange Park, Ill.: 1979.

Burnham, David. "How National Academy of Science Decided to Halt a Nuclear Waste Report Is Disputed." *The New York Times,* New York: June 25, 1979.

Caldicott, Helen. *Nuclear Madness: What You Can Do!* Brookline, Ma.: Autumn Press, Inc., 1978.

Campbell, Marrie, and Mike Kirk. "Do I Look Like I Want to Die?" KCTS Television. Seattle: June 25, 1979.

Carter, Luther J. "Academy Squabbles over Radwaste Report." *Science,* Vol. 205 (July 20, 1979), pp 287–89.

Cohen, Bernard L. "The Disposal of Radioactive Wastes from Fission Reactors." *Scientific American,* Vol. 236 (June 1977), pp. 21–31.

———. and I-Sing Lee. "A Catalog of Risks." *Health Physics,* Vol. 36 (June 1979), pp. 707-22.

Commoner, Barry. *The Politics of Energy.* New York: Alfred A. Knopf, 1979.

Comparative Risks of Different Methods of Generating Electricity, The. American Nuclear Society, La Grange, Ill.: October 1979.

Crowell, Todd. "A Basalt Cemetery for Nuclear Waste." *New Scientist* (February 21, 1980), p. 548.

"Danger at Three Mile Island." CBS Television Network, New York: March 30, 1979.

Deutsch, R. W. *Nuclear Power.* General Physics Corporation, Columbia, Md.: 1979.

Fermi, Laura. *The Story of Atomic Energy.* New York: Random House, 1961.

Ford, Daniel F. *A History of Federal Nuclear Safety Assessments: from Wash-740 Through the Reactor Safety Study.* Cambridge: Union of Concerned Scientists, 1977.

Haglund, Keith. "Nuclear Power and Health." *New Physician* (June 1979).

Holden, Constance. "NRC Shuts Down Submarine Fuel Plant." *Science.* Vol. 206 (October 5, 1979), pp. 30-32.

Joyce, Chris. "Carter Reveals Nuclear Waste Plans." *New Scientist* (February 21, 1980), p. 548.

Lilienthal, David E. *Atomic Energy: A New Start.* New York: Harper & Row, 1980.

Lovins, Amory B. *Soft Energy Paths.* New York: Harper & Row, 1979.

McCracken, Samuel. "The Harrisburg Syndrome." *Commentary* (June 1979).

McPhee, John. *The Curve of Binding Energy.* New York: Ballantine Books, 1974.

Magnetic Fusion. U.S. Department of Energy, Washington, D.C.: April 1978.

Marcus, Steven J. "The Risks of Risk Assessment." *Technology Review* (May 1979), p. 82.

Marshall, Eliot. "A Preliminary Report on Three Mile Island." *Science.* Vol. 204 (April 20, 1979), pp. 280-81.

——. "Carter Backs 'Spirit' of Kemeny Report." *Science.* Vol. 206 (December 21, 1979), p. 1380.

Miller, Marvin. "The Nuclear Dilemma: Power, Proliferation, and Development." *Technology Review* (May 1979), pp. 18-29.

Nader, Ralph, and John Abbotts. *The Menace of Atomic Energy.* New York: W.W. Norton & Company, 1979.

Nuclear Power and the Environment. American Nuclear Society, La Grange Park, Ill.: 1976.

Nuclear Power: Answers to Your Questions, rev. ed. Edison Electric Institute. Washington, D.C.: 1979.

"Nuclear Power: Can We Live with It?" *Technology Review* (June–July 1979), pp. 32–47.

Nuclear Power Quick Reference II. General Electric Nuclear Energy Group, San José, Cal.

"Real News from Three Mile Island, The, Interview with Russell Schweickert." *The CoEvolution Quarterly* (Summer 1979), pp. 2–10.

Solar Energy. U.S. Department of Energy, Washington, D.C.: March 1978.

Stevens, Ted. "Leaky Defence of Nuclear Waste." *New Scientist* (July 31, 1980), p. 349.

Stobaugh, Robert, and Daniel Yergin, eds. *Energy Future.* New York: Random House, 1979.

Torrey, Lee. "The Week They Almost Lost Pennsylvania." *New Scientist* (April 19, 1979), pp. 174–78.

Tye, Lawrence S. *Looking but Not Seeing.* Union of Concerned Scientists, Cambridge, Ma.: 1979.

Union of Concerned Scientists. *The Nuclear Fuel Cycle.* Cambridge, Ma.: Massachusetts Institute of Technology Press, 1977.

Annual Report, 1978. U.S. Nuclear Regulatory Commission, Washington, D.C.

"U.S. Reviews Nuclear Waste Disposal . . . and So Does West Germany." *New Scientist* (October 26, 1978), p. 253.

Uranium: Energy for the Future. Atomic Industrial Forum, Inc., Washington, D.C.: 1978.

"Wasteful Truth About Soviet Nuclear Disaster, The." *New Scientist* (January 10, 1980), p. 61.

Wells, H. G. *The World Set Free.* New York: E. P. Dutton, 1914.

What You Should Know About the Hazards of Nuclear Power. Union Of Concerned Scientists, Cambridge, Ma.

Wilson, Richard. "Analyzing the Daily Risks of Life." *Technology Review* (February 1979), pp. 41–46.

INDEX

Advertising, pronuclear, 117–118, 133, 151
ALAP ("as low as practicable") radioactive discharge, 37–38
Alpha rays, 15, 16, 33, 47, 55, 120
Antinuclear movement, 29–30, 115, 116, 117, 121, 134, 135–139, 142
Atomic bomb, 19, 20, 95
 student's construction of, 90
Atomic Energy Act:
 of 1946, 93, 95
 of 1954, 95
Atomic Energy Commission (AEC), 61, 93, 95–96, 97, 98, 99, 100, 101, 105, 111, 122, 123, 124, 125, 127
 abolition of, 100
 and Advisory Committee on Reactor Safeguards, 98
Atomic Industrial Forum (AIF), 27, 114, 120, 137, 143
Atomic pile, test of, 19, 23
Atoms, 13–14, 17–19

Background radioactivity, 31–32, 34, 36, 119
Becquerel, Antoine Henri, 15
Beta rays, 15, 16, 33, 53
Boric acid solution, 23, 24, 55
Boron rods, 23
Breeder reactor, 29, 103, 116, 138, 141
Brown's Ferry (Ala.) nuclear power plant, 110
 fire at, 83–84, 98, 108

Cadmium, 19, 23
Caldicott, Helen, 113, 114, 120, 123, 134
Canada, 47, 86, 87, 141, 142
Cancer:
 causes of, other than radiation, 36
 nuclear energy for fighting, 20
 radiation-induced, 35, 36, 48, 59, 113, 114, 119, 120
Carbon, 14, 32
Carbon-14, 39
Carter, Jimmy, 58, 87
Cesium-137, 37, 40
Chain reaction, nuclear, 18, 19, 23
China, 20, 142
China Syndrome, The (film), 116, 121
Coal, 25, 118

156

Coalition for a Non-Nuclear World, 135, 142
Committee of Energy Awareness, 133
Consolidated Edison Company, 85, 96, 97, 110
Coolant for reactor, 23
 and LOCA, 69–70, 77
Cosmic rays, 31
Cost-plus, definition of, 107
Cover-ups within nuclear establishment, 126–127, 130, 132
Curie, definition of, 33
Curie, Irène, 33
Curie, Marie, 15, 33
Curie, Pierre, 15, 33

Defense Department, 105
Defense-in-depth, and ECCS, 70–74, 77
Department of Energy (DOE), 27, 51, 101, 105, 124, 125, 126, 144, 145
Deutsch, Robert W., 116, 117, 131
Dumps, nuclear, 60–61

Edison Electric Institute, 119, 121
Eisenhower, Dwight D., 122, 123
Electricity:
 generated by burning wood, 144
 generated by nuclear energy, 21, 25, 27, 28, 29
Electrons, 14, 33, 55
Elements:
 half-life of radioactive, 16
 isotopes of, 14–15
 radioactive, 15, 16
 transmutation of (radioactive decay), 16, 17
Emergency Core Cooling System (ECCS), 98
 at Brown's Ferry plant, 84
 and defense-in-depth, 70–74, 77
 tests on, 74–76
 at Three Mile Island, 77, 79, 80, 81, 91
Energy Reorganization Act (1974), 100

Energy Research and Development Administration (ERDA), 101, 124, 125
England, 20, 141
Environmental Protection Agency (EPA), 34, 35

Fermi, Enrico, 17, 18, 19, 23, 96
Fermi reactor (Mich.), accident at, 85
Fission, 18, 19, 23, 24, 25, 28, 29, 37, 41, 53, 58, 149
Fluorine, radioactive, half-life of, 16
Food chain, radioactivity in, 39–40, 50–51
Fossil fuels vs. uranium, 25–28
France, 20, 141, 142
Fusion, 148–149

Gamma rays, 15, 16, 32, 33, 53
Geothermal energy, 145
Germany, West, 141, 142
Graphite, and fission, 18, 23
Guayule, 144–145

Half-life of radioactive element, 16
Hard energy, 149
 vs. soft energy, 146–148
Helium, 14, 148
Hiroshima, 19
Hydrogen, 14
 and fusion, 148, 149
 heavy, 14
 isotopes of, 14–15

India, 86, 87
Indian Point (N.Y.) nuclear power plant, 96, 97
 accident at, 85, 110
 inadequate inspection of, 99–100
 inadequate security at, 89
Inhaber, Herbert, 140, 141, 147, 149
Iodine-131, 37, 39
Ionization, 33
Iron-59, 37, 39, 40
Isotopes, 14–15
 radioactive, 32, 39, 40, 55, 59

Index **157**

Japan, 19, 71, 90, 141, 142, 143
Jungk, Robert, 117

Kemeny, John G., 80, 81, 91
Kemeny Commission, 82, 92, 93, 101, 102, 130
Kilowatt-hour, definition of, 27
Krypton-85, 37, 39, 40

Lee, I-Sing, 139
Leukemia, 35
LOCA (loss-of-coolant accident), 69-70, 77
Lung cancer:
    plutonium as cause of, 59, 113, 114, 120
    in uranium mines, 48

Magnetic Fusion Engineering Act (1980), 148-149
Mancuso, Thomas F., 124, 125
Massachusetts Public Interest Research Group (MPIRG), 45
McCracken, Samuel, 116, 117
Media, and nuclear news, 130-132
Meltdown, 70, 77, 87, 111, 128, 131
Metropolitan Edison Company, 80, 81, 106, 109, 110, 112, 128, 129, 130
Mexico, 141, 142
Middle East, tensions in, 27
Milling and mill tailings in nuclear fuel cycle, 48-52
Mobil Oil Corporation, 117, 118

Nader, Ralph, 112, 120, 127
Nagasaki, atomic bomb dropped on, 19
National Academy of Sciences (NAS), 125-126, 147, 148, 149
National Aeronautics and Space Administration (NASA), 144
National Reactor Testing Station, 74, 105
Neutrons, 14, 17
    as bullets for splitting atoms, 17, 18, 19

Nevada, nuclear weapons tested in, 122, 123
New York Public Interest Research Group, 65
News media, and nuclear establishment, 130-132
Nitric acid, fuel rods immersed in, 55
North Anna (Va.) nuclear power plant, 97, 98, 100
Nuclear chain reaction, 18, 19, 23
Nuclear dumps, 60-61
Nuclear energy:
    alternatives to, 144-146, 149
    costs of, 107, 118
    vs. fossil fuels, 25-28
    future of, 134, 141-144, 147, 150-152
    and taxpayers, 107, 118
    uses of, 20-21
    See also Nuclear industry; Nuclear power plant(s)
Nuclear fuel cycle, 46, 47 ff.
    back end of, 47, 53-55, 56-57, 118
    enrichment step in, 53, 106
    fabrication step in, 53
    front end of, 47, 52, 53
    and gaseous uranium, 52-53, 106
    milling and mill tailings in, 48-52
    radioactivity released throughout, 117-118
    and reprocessing, 55, 58-59, 99, 102, 106, 118, 143
    See also Nuclear power plant(s); Nuclear reactor
Nuclear industry, 104
    advertising by, 117-118, 133, 151
    cover-ups within, 126-127, 130, 132
    deception within, 127-130
    government aid to, 105-106
    and news media, 130-132
    and public relations, 118-120, 121, 132, 133, 151
    See also Nuclear energy; Nuclear power plant(s)
Nuclear power plant(s), 21-25

Reagan, Ronald, 138
Rems (roentgen equivalent in man), 33, 34, 35, 38, 41, 43, 44, 45
Reprocessing, 55, 58–59, 99, 102, 106, 118, 143
Richland (Wash.), radiation study at, 124, 125
Rio Puerco (N.M.), mill-tailings spill into, 106
Risk analysis, risks of, 139–141
Roentgen, definition of, 33–34
Roentgen, Wilhelm, 33–34
Rutherford, Ernest, 14, 15, 16

Sabotage, nuclear, 88, 90
Seaborg, Glenn T., 96, 125
Seabrook (N.H.) nuclear power plant, 135, 151
Soft energy, 145, 146, 149, 150
vs. hard energy, 146–148
Solar power, 140, 145, 146, 147, 148, 149
Soviet Union, 20, 47, 126, 127, 148
Strontium-90, 37, 39, 40
Subatomic particles, 14
Sweden, 141, 142–143
Synroc (synthetic rock), and radwastes, 62
Synthetic fuels, 145, 149
Szilard, Leo, 11, 12, 17, 18

Technology Review, 141
Television color set, radiation emitted by, 32
Tennessee Valley Authority (TVA), 83, 108, 109
Terrorism, nuclear, 87–88, 90, 102
Theft, nuclear, 87, 89–90, 106
Thorium, 48, 117
Thornburgh, Richard, 129, 131
Three Mile Island:
 accident at, 12, 40, 66–67, 76–80, 83, 91, 100, 102, 108, 111, 128
 cleanup bill for, 106, 109, 110
 and investigation, 80–82
 news coverage of, 130–131

Ultraviolet light, 31
Union of Concerned Scientists (UCS), 45, 62, 63, 64, 76, 86, 102
Uranium, 14, 15, 18, 19, 23, 32, 36, 99, 117, 118, 136, 146
 vs. fossil fuels, 25–28
 gaseous, at enrichment plants, 53, 118
 half-life of, 16
 isotopes of, 15, 28, 29
 and milling and mill tailings, 48–52
 mining of, 47–48, 136
 radioactive decay of, 16
 reserves of, 28, 29
Uranium Mill Tailings Radiation Control Act (1978), 51
Uranium-235, 28–29, 41, 53, 58, 69, 118, 146
Uranium-238, 28, 29, 37, 53
Uranium-239, 29

Virginia Electric & Power Company (VEPCO), 97, 100

Wastes, nuclear: *see* Radwastes
Water power, 144, 145, 146, 149
Wells, H. G., 9, 10, 11, 12, 29
West Valley (N.Y.) reprocessing plant, 106
Wind power, 140, 144, 145, 146, 147, 149
World Set Free, The (Wells), 9, 10, 11, 17
World War II, 18, 19, 90

X rays, 16, 32, 34

Yellowcake, 48, 50, 52, 53

Zion (Ill.) nuclear reactors, 97
Zircalloy fuel rods, 23, 53

160 *Index*

accidents at, 81, 83-86 (*see also* Three Mile Island)
containment of radioactivity within, 36-39
costs of, 107, 108, 118, 152
inspection of, inadequate, 99-100
jumpers employed at, 43-45
and LOCA, 69-70, 77
moratorium on, unofficial, 152
number of, 38, 39
pollution problems with, 25
protests against, *see* Antinuclear movement
radioactive release by, 30, 37, 38, 39
safety systems for, 67-74
security at, inadequate, 88-89
shutdown of, 41, 42, 108
workers at, protection for, 41-43
See *also* Nuclear energy; Nuclear fuel cycle; Nuclear industry; Nuclear reactor; Radwastes
Nuclear proliferation, 86-87, 103
Nuclear reactor, 23, 29, 37, 41
disposal of worn-out, 64-65
See *also* Nuclear fuel cycle; Nuclear power plant(s)
Nuclear Regulatory Commission (NRC), 34, 35, 37, 41, 43, 44, 49, 51, 71, 75, 76, 79, 81, 92, 93, 101, 102, 125, 128, 129, 130, 131
Nuclear terrorism, 87-88, 90, 102
Nuclear theft, 87, 89-90, 106

Oak Ridge National Laboratory (Tenn.), 74, 105, 124, 125
Ocean areas, as warm-water energy source, 145
Oil:
costs of, 27, 118
U.S. imports of, 26-27
U.S. production of, 26
uses of, 28
Ozone, 31

Pacific Gas and Electric Company, 145

Particles, subatomic, 14
Pigford, Thomas, 80, 81
Plutonium, 29, 37, 55, 58, 59, 63, 87, 96, 113, 114, 115, 116, 120
Pollard, Robert, 102
Polonium, 15, 16, 32
Portsmouth Naval Shipyard (N.H.), 123
Price-Anderson Act, 111, 112, 119, 120, 136
Pronuclear advertising, 117-118, 133, 151
Pronukers, and war of words, 115, 116
Protons, 14, 33
positive charge of, 14
Public Utilities Commission (PUC), 107, 108, 110

Rad (radiation absorbed dose), 33, 34
Radioactive decay, 16, 17, 32
Radioactivity, 15, 16, 119, 149
background, 31-32, 34, 36, 119
and disease, 33, 35, 36
and environment, 39-40, 50-51
in food chain, 39-40, 50-51
low-level exposures to, 35-36, 40, 119, 123-124
manmade, 30, 32, 34, 36
released throughout nuclear fuel cycle, 117-118
and safety standards, 33-35
Radium, 15, 16, 32, 47, 48
Radwastes, 59, 118, 132
and disposal of worn-out reactors, 64-65
dumps for 60-61
low-level, 60
and NAS, 125-126
problems in storage of, 59, 62, 63, 64, 99, 103, 106, 118
proposals for storage of, 62
Rasmussen, Norman C., 75, 76
Rasmussen Report (Reactor Safety Study), 75-76, 127
Reactor, nuclear: *see* Nuclear reactor
Reactor Safety Study (Rasmussen Report), 75-76, 127

Index **159**

333.79 Weiss, Ann E.
WEI
   The nuclear
      question

		DATE	

© THE BAKER & TAYLOR CO.